FOR THE LOVE OF

Pennsylvania's Breweries

DR. ALISON E. FEENEY

FOR THE LOVE OF BEER: PENNSYLVANIA'S BREWERIES

1405 SW 6th Avenue • Ocala, Florida 34471 • Phone 352-622-1825 • Fax 352-622-1875
Website: www.atlantic-pub.com • Email: sales@atlantic-pub.com
SAN Number: 268-1250

Library of Congress Cataloging-in-Publication Data

Names: Feeney, Alison, 1969- author.
Title: For the love of beer : Pennsylvania's breweries / by Alison Feeney.
Description: Ocala, Florida : Atlantic Publishing Group, Inc., [2018] | Includes bibliographical references and index.
Identifiers: LCCN 2018001799 (print) | LCCN 2018002783 (ebook) | ISBN 9781620235119 (ebook) | ISBN 9781620235102 (pbk. : alk. paper) | ISBN 1620235102 (alk. paper)
Subjects: LCSH: Breweries—Pennsylvania. | Beer—Pennsylvania.
Classification: LCC TP573.P46 (ebook) | LCC TP573.P46 F44 2018 (print) | DDC 663/.309748—dc23
LC record available at https://lccn.loc.gov/2018001799

Printed in the United States

PROJECT MANAGER: Danielle Lieneman
INTERIOR LAYOUT AND JACKET DESIGN: Nicole Sturk

Over the years, we have adopted a number of dogs from rescues and shelters. First there was Bear and after he passed, Ginger and Scout. Now, we have Kira, another rescue. They have brought immense joy and love not just into our lives, but into the lives of all who met them.

We want you to know a portion of the profits of this book will be donated in Bear, Ginger and Scout's memory to local animal shelters, parks, conservation organizations, and other individuals and nonprofit organizations in need of assistance.

– Douglas & Sherri Brown,
President & Vice-President of Atlantic Publishing

Logan, Hudson, Sterling, and Matt:
Thanks for all the love, support, and endless road trips

Acknowledgments

I have sipped many beers with many people along the way who all made this book possible. However small and trivial that may seem, I greatly value those times and conversations. Beer has a tendency to break down barriers and encourage people to sit and talk to one another. From strangers at bars, to my tennis teammates, to some of my closest friends, and of course to my family, those conversations helped develop many of the ideas presented in this book.

I must thank my Department Chair, Dr. William Blewett, and my Dean of the College of Arts and Sciences, Dr. James Mike, who supported my academic research that lays the foundation for each chapter. I need to thank Dr. Joe Poracsky, not only for my early training and many afternoons in Portland's brewpubs, but for a simple email entitled "Joe's 2 Cents" that has guided me during my professional career to ensure that I love what I do and pursue those endless hours of work with passion. Atlantic Publishing has been amazing to work with, particularly their Publisher Consultant, Jack Bussell, and my Project Manager, Danielle Lieneman. Of course, I need to thank Matthew Fetzer who is my field researcher, designated driver, and partner. And most of all, the hop farmers, malters, and brewers that work tirelessly to provide us with excellent beer.

Cheers to all!

Contents

Valley Brewing Company, Barley Creek Brewing Company, Bullfrog Brewery, Elk Creek Café & Aleworks, Neshaminy Creek Brewing Company

Featured Breweries: Tröegs Brewing Company, Turkey Hill Brewing Company, The Vineyard and Brewery at Hershey, Moo-Duck Brewery, Wyndridge Farm, Center Square Brewing, Big Bottom Brewery, Stable 12 Brewing Company, Root Down Brewing Company, Aldus Brewing Company, Miscreation Brewing Company, Something Wicked Brewing Company

Featured Breweries: Burd's Nest Brewing Company, Desperate Times Brewery, Harty Brewing Company, Red Castle Brewery & Brewpub, ShawneeCraft Brewing Company, Pocono Brewery Company, North Slope Brewing Company, Yorkholo Brewing Company, The Wellsboro House Brewery, Clarion River Brewing Company, Mortals Key Brewing Company, Rusty Rail Brewing Company, Mount Gretna Craft Brewery, Columbia Kettle Works, Conshohocken Brewing Company, GearHouse Brewing Company, Ever Grain Brewing Company

Featured Breweries: Yards Brewing Company, Fegley's Brew Works, Snitz Creek Brewery, Roy Pitz Barrel House, Free Will Brewing Company, Rhone Brew Company, Bonn Place Brewing Company

Introduction

People love beer. It is the third most widely consumed drink in the world after water and tea.[1] People have always loved beer. It has been argued that people loved beer so much it was the impetus for leaving a nomadic life behind and becoming a settled agricultural society in order to grow crops needed for beer.[2] It inspired people to make technological innovations and scientific advancements in order to make better beer, store beer longer, and transport beer farther. Beer was one of the first religious offerings and remains a component of many social customs throughout the world. It was a standard traded commodity before we had money, and it remains a globally important industry.

Today, Americans consume over 20 gallons of beer a year,[3] and although large mass-produced beer still holds the market, many Americans are selecting locally made craft beers instead. Prior to Prohibition, Pennsylvania led the nation in the number of small independent breweries because of its diverse immigrants and ideal geography. In the past few decades, the number of craft breweries in Pennsylvania has exploded.

1. Patterson M. and N. Hoalst-Pullen. 2014.
2. Kohn, Rita T. 2010.
3. Reid, Neil, Ralph B. McLaughlin, and Michael S. Moore. 2014.

PENNSYLVANIA BREWERIES

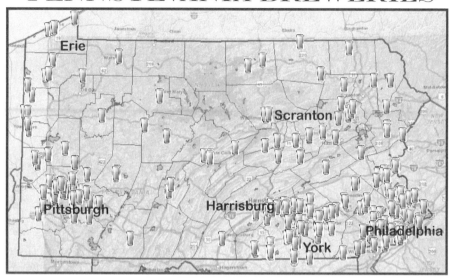

The growth in the quality and quantity of these breweries has been steady. They seem to follow a similar business model: renovate an older building, decorate with antiques and local artifacts, and cater to a population that seeks out local attachments and deliberately avoids chain stores and super-sized global brands. Rarely do you enter a brewery without seeing one of the owners and/or brew masters still hard at work. Despite the numerous long hours, they still remain anxious to talk to their customers, pride themselves in creative flavors, and stress their passion for their craft. This male-dominated field, with bartenders and brew masters usually sporting several days of unshaven facial hair, have a laid back disposition, do what they love, and love what they do. They create a love of beer, catering to the true beer connoisseurs, while inspiring others to pick up home brewing and encouraging the average beer drinker to be more adventurous and try something outside of their "Bud Light comfort zone."

Like many of you, I too love beer. I'm not the typical craft brewer or even the typical craft beer consumer. I'm not that young male sporting the trendy lumbersexual beard but rather a petite female with naturally-born bimbo blonde hair. But behind this facade of someone that looks more like

a stereotypical woman in a flavorless light beer advertisement, I have an insatiable desire for exploration and adventure that has led me to brew houses throughout North America. In doing so, I have seen what makes an area unique. The simple enjoyment of a beer in the cozy confines of a local brewpub moves people to take pictures, post comments on social media, and promote that special something that local people are proud to call their own.

I started graduate school in Portland, Oregon in the early 1990s. Many great afternoons were spent discussing research in the city's new exciting craft brew pubs with my advisor and fellow grad students. I fell in love with these urban confines that served different and unique flavored beers. But as much as I enjoyed the beverages, I was falling in love with the culture that surrounds the entire industry. Beer has a fascinating social history. A beverage that was born out of dietary necessity evolved into one that was considered part of early North American women's domestic duties and later into one that represented the culture and diversity of its illustrious immigrants. Over the past 100 years, it became a banned product during Prohibition to an icon of the industrial mass produced era that still takes center stage in the advertising of our sports and entertainment industry today. In the past few decades, beer has become this alluring treat that people actively seek out, and they travel those extra miles to experience new brews and seasonal flavors' at its place of origin. Craft beer enthusiasts pilgrim to highly revered breweries, hunt down new brews in remote towns, and venture into different neighborhoods within larger cities, recording and posting their findings on phone apps and sharing their experiences on social media with the world.

Local beers do taste different. Their foundation is built on different grains, often adding flavors from local farm products and culminating with different minerals originating from the local watershed. Often the quality of the watershed is marketed with pristine images portrayed in the branding of many beers. Local geography is often epitomized in craft beers with the creative names that embody and commemorate the local environment, historic past, hometown heroes, or colorful legends.

I moved from the Pacific Northwest — surrounded by free loving liberals with strong environmental convictions — to the Great Lakes and the industrial heartland, where I grew to love the enduring spirit of the blue-collar worker that built this nation. Milwaukee, Detroit, Toledo, Minneapolis, and Chicago are just a few of the great beer-drinking cities where immigrants have left a vibrant mark on the cultural identity on this county. Nowhere is this human footprint and resolve move evident than watching someone swing a terrible towel at a Pittsburgh Steelers' game. "Yinz" know what I'm talking about. Before and after the games, the local dialect that is derived from Scotch-Irish and German influences with a rhythmic pattern similar to those found in Croatian, Slavic, and other Eastern European languages is heard in the numerous renovated bars throughout the Golden Triangle. Craft breweries have been and remain an integral part in building and reviving downtown Pittsburgh.

I now live and work in Pennsylvania and am proud to call it home. I try to experience as much of the state's vast physical, historical, and cultural resources as possible, with of course, stops at local breweries. I have found that Pennsylvania's brewers exhibit an industrial spirit and an entrepreneurial passion. They are truly committed to their craft of brewing great beer, fiddling with recipes and techniques, and although people may enjoy different flavors or palates, it is undeniable that attentive details were laboriously expended in wielding their product.

The dedication brewers have to their craft extends well past the brewpub and customer service. I was fortunate to observe this sincerity and commitment one Sunday afternoon in the beautiful Cumberland Valley countryside. The brewers of **Molly Pitcher Brewing Company** met with the owners of Sunny Brae Hops. Adam and Diana Dellinger grow high quality hops that they continuously monitor throughout the growing season. Using their expertise in soil, agriculture, and geo-environmental science, they are actively engaged in growing and providing the best possible products for local brewers. Because the different types of hops can vary based upon minor difference in soils and climatic conditions, Sunny Brae meets with the brewers, describes the year's harvest, and educates their customers on exactly what they are purchasing.

The owners of **Molly Pitcher Brewing Company** were excited to purchase local hops and were completely dedicated to supporting their local farm. Not only did they purchase the product, but they also helped harvest the crop. It was a collaborative effort operating tractors, cutting bines, feeding the hop machine, and picking quality flowers.

Typically craft brewers work with hops that are converted and stored as dry pellets; however, in the fall brewers throughout the state compete for the limited supply of fresh, locally grown Pennsylvania hops. They are eager to create special aromatic beers made with the wet hops that take full advantage of the resins and oils from the freshly picked flowers. **Molly Pitcher Brewing Company** transported the hops from the farm to their brewery and within hours of being picked from the bine, they started the fermentation process of their Cascade Wet Hops. They recently opened a taproom on Carlisle's High Street, in the vibrant historic district. By opening a second taproom, the new spacious two-story pub on High Street can accommodate many more customers, and it opened up more space in the original brewery on South Street, allowing them to increase production.

Carlisle was settled in 1751, and the town has beautifully preserved many colonial era historic buildings. The town remains steeped with military history, as it is home to the U.S. Army War College and the U.S. Army Heritage and Education Center. **Molly Pitcher Brewing Company** cele-

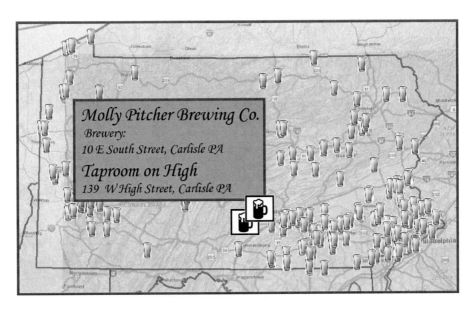

Molly Pitcher Brewing Co.

Brewery:
10 E South Street, Carlisle PA

Taproom on High
139 W High Street, Carlisle PA

brates Carlisle's most heroic female resident who fought in the American Revolutionary War. Legend has it, that as the Battle of Monmouth ensued she carried pitchers of water to the soldiers, however when her husband was wounded and carried off the battlefield she took his place loading the cannons. After the battle, General Washington commemorated her courage and issued her as a noncommissioned officer. To celebrate the local history, the brewery has creative tap handles made with antique wooden muskets to pour their flagship beers with Revolutionary War inspired names: Independence IPA, Redcoat ESB, Black Powder Stout, and Cannonball Kolsch.

Collectively, brewers are dedicated to building stronger communities. With their re-investments or reuse of older buildings, breweries aid in the development of neighborhoods and towns. They try to support other local businesses by purchasing products from nearby farms and endorse local craftsmen with their interior designs. Many breweries overtly advertise their sustainable business models on their website and social media. They demonstrate their commitment to recycling and composting programs, creating equitable work conditions for employees, and some have even absorbed additional expenses by promoting environmental efforts and using solar and wind powered energy. On any given night, this commitment to the community is apparent. Most breweries have created a communal tavern with a relaxed family atmosphere for friends and neighbors to gather, with sounds of local musicians filling the rooms and advertisements supporting local charity events displayed on the walls.

Pennsylvania has a long, rich beer-drinking culture that is closely embedded to its historic, social, and economic past. Our Founding Fathers established a brewery as one of the first buildings constructed in every new town they settled, and our Founding Mothers cooked a mean brew in every home in the rural countryside. The abundant rivers and access to open oceans made the physical geography ideal for the influx of beer-drinking European immigrants to make Pennsylvania the nation's largest producer and consumer of beer for several hundred years. This strong foundation allowed for several breweries to survive Prohibition, and they proudly display their longevity on their marketing campaigns.

Pennsylvania now leads the nation in barrels of production, with a $5.8 billion economic impact to the state [4]. Breweries across Pennsylvania provide jobs, renovate urban areas, create tourism regions, and develop cultural events with undeniably significant impacts.

The point of this book is not to inventory every brewery in Pennsylvania. Although some listings, addresses, and many maps do exist throughout this book, the intent isn't to be an all-inclusive encyclopedia. The industry is too dynamic, and the book would be obsolete before it is published. Rather, this book is intended to provide a sense of brewing history and Pennsylvania's connections between beer and its geographic wonders and cultural richness. It is intended to explain some of the stories behind a few of the most enjoyable beers while detailing the unique buildings and architectural treasures that are renovating urban areas and reviving communities. Whether you are an amber ale aficionado, dunkelweizen drinker, sour snob, saison sipper, porter pounder, IPA imbiber, chocolate-coffee connoisseur, quadruple-Belgian quaffer, German-pilsner guzzler, or simply en-

4. Brewers Association. 2017.

joy a swig of soda, this book is an ongoing statement to the dedication of Pennsylvania's craft brewers and the LOVE OF BEER.

CHAPTER 1

History of Beer

Posted over the men's urinal at my local brewery is a small sign that reads "Beer: helping ugly guys get laid for over 5,000 years." Apparently men have been desperately hanging on to this belief since the evolution from pictorial to a formal written language. A small clay tablet — created in Mesopotamia around 1800 BC and now held at the British Museum in London — depicts Old Babylonian culture. The plaque shows a man and woman having sex, while the woman bends over to drink beer through a straw. This molded and fired plaque was mass produced. Although experts are unsure of the purpose of these plaques and have suggested the possibility of it having magical or religious significance, clearly beer is central to the theme of erotic pleasures. According to the Epic of Gilgamesh — written around this same time period and often regarded as the first great piece of literature — beer drinking, along with sex, food, and oil-rubs are fundamental elements to what makes us human.

Biologically, beer might not be the correct genetic description of what makes us human, but it appears that from the earliest beginnings in agricultural communities to the presences of primetime advertising slots during the Super Bowl, beer has held a vital role in human history and remains one of the most widely consumed beverages in the world, after water and tea.

More than 12,000 years ago, beer was first brewed in Mesopotamia. It was probably by accident and most likely tasted horrible, but it must have had some redeeming qualities because people made it again. And again. And again. Enough so, that many archaeologists argue that it was beer, not food, that led to the domestication of plants, and it was the love of beer that motivated people to stop hunting and gathering to settle in one place in order to tend crops.[5] Beer was made from any plant that could transform starch to sugars, but barley was used most often and may have been our ancestor's first cultivated crop. That love of beer and the cultivation of barely led to great inventions that changed the world forever. Arguably, the great technological developments such as the wheel, plow, irrigation, writing, chemistry, mathematics, and even political institutions were all created to help humans make more beer.

Increasingly, archaeological, ethnographic, and textual artifacts from around the world demonstrate humans' ability to ferment any available

5. Homan, Michael M. 2004.

sugars into alcoholic beverages.[6] Only the extreme Polar Regions — where geography limits production — did the Eskimos in the north, the people of Tierra del Fuego in the south, and the Australian aborigines live without alcohol. Residue on 4,000 BCE ceramics, found in the Zagros Mountains, holds the earliest known chemical evidence of beer production, but numerous artifacts dating to much earlier times suggest grains and water were allowed to naturally ferment. In the Ancient Near East, scholars focused more on wine and downplayed the importance of beer, yet the accumulation of evidence indicates that beer was a dietary staple, and numerous surviving artifacts support its popularity. Pottery vessels, stoves, and underground taverns found in Mijiaya, China are laden with evidence that provide ancient beer recipes of broomcorn millet, barely, and other tubers and demonstrate the advanced knowledge and beer-making techniques that these ancient peoples had to brew and store beer. Archaeological excavations reveal that many Chinese temples contained a brewery, and their surviving pictorial and written literature is enriched with statements of gods and deities linked to beer drinking or beer production. Beer may have been an integral piece in the changing complexities of societal relations during the time. This social importance, along with applied beer-producing skills, may have been a motivating factor in initial translocation and dispersion of barley from Western Eurasia to become a subsistence agricultural crop in the Central Plains of China nearly 3,000 years later.

What is often overlooked in history is the fact that until very recently, women were the driving force behind much of the world's beer production. While men were out hunting and gathering the women were in charge of brewing beer. As humans transitioned from a nomadic lifestyle to organized civilizations, the women's domain included production, storage, and distribution of beer. They often served as tavern keepers and were the economic foundation of many early urban societies. Not surprisingly, many early peoples paid tribute to women brewers and had highly respected beer goddesses. The hymn to Ninkasi, the Sumerian goddess, is effectively a recipe for making beer. Another Sumerian goddess, Siris watched over the

---

6. McGovern, Patrick E.,et al. 2005.

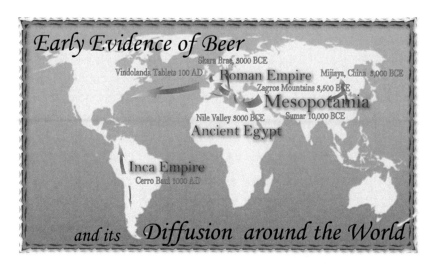

daily brewing of beer, and Mbaba Mwana Waresa, the Zulu fertility goddess, is beloved for her invention of beer.

In Pennsylvania, a woman continued this tradition and was a driving force in the development of the state's craft brewery resurgence. Carol Stoudt opened **Stoudt's Brewing Company** in 1987 as the first female brew master in the United States since Prohibition. Often regarded as the Queen of Hops, she used her background in chemistry and microbiology to develop her brewery. She nicely integrated the brewery into a larger complex in Adamstown that includes the Black Angus restaurant and a market that sells artesian cheeses, breads, and antiques. They are all committed to craftsmanship and pride themselves on local identity and local ingredients. .

The bar and restaurant decor proudly displays the evolution of the establishment over the past 50 years.

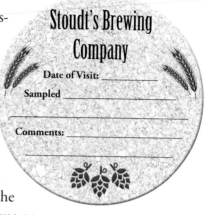

Stoudt's Brewing Company

Date of Visit: _____

Sampled _____

_____

_____

Comments: _____

_____

Humans all around the world and throughout history leave behind similar evidence of their love of beer. One of the world's earliest known breweries lies in the remains of Skara Brae, a Neolithic settlement found on the Orkney Islands, Scotland.[7] The village was occupied between 3180 and 2500 BCE, where malted grains were brewed with local herbs to induce a trance during early rituals. Chemical results from late Neolithic pottery found along the Yellow River, China, dating to approximately 2400 and 2200 BCE, confirm the existence of fermented beverages.[8] The early inhabitants of this region not only relied on these beverages as a dietary staple, but the drinks contributed significantly to their social life and were a major part of ceremonial offerings and feasting activities.

Beer is integral to cultures on many different social, religious, economic, and political levels. Beer and its social connotations have transcended human customs over the millenniums. In early non-monetary societies, wages, pensions, and taxes were commonly paid in beer. Throughout the ancient world, religious temples and administrative buildings were all built by paying their workers in bread and beer, and most of the buildings they constructed contained a brewery.[9] Of course to ensure fair pay, standard measures were set. It is not surprising that one of the earliest and most complete listing of laws, Hammurabi's code, dictates the price and strength of beer.

---

7. Dineley, M., & Dineley, G. 2000.
8. McGovern, et al. 2004.
9. Godlaski, Theodore M. 2011.

Rumspringa Brewing Company, located in Lancaster County between Intercourse and Bird-in-Hand, cites Hammurabi's code on their walls. The name of the brewery comes from the age-old tradition where the local Amish youth are given a temporary period to explore activities outside of the Amish community before they officially join their church. In honor of this local culture, **Rumspringa Brewing Company** invites people to try "Sow Your Wild Oats" in the upstairs tasting room along with German style food. Visitors can enjoy sampling Mt. Hope wines and ciders and shop for kitchen and gourmet accessories as they pass through the first floor of the barn.

Shenas were bakeries and breweries found in every ancient Egyptian temple. In addition to being the central location of food production, they served as the primary hub for all economic, administrative, and spiritual activities. Beer was the staple for both the living and the deceased. It was the source of daily rations and payments in the actual world and as spiritual offerings in the afterworld. Ancient Egyptians believed that deceased pharaohs' spirits communicated the *maat*, the law of the universe, with the living in return for nourishment, prayers, and gifts.[10]

Rumspringa
Brewing Company

Date of Visit: _____

Sampled _____

_____

Comments: _____

10. Smith, Vanessa. 2006.

Sharing beer linked the mortals with gods, the commoners with kings, and connected those living in the present to those who had lived in the past.

The Roman Empire also paid soldiers in beer. The Vindolanda tablets are the oldest surviving written records in Great Britain and provide insights into the lives of Roman soldiers stationed along the northern frontier near Hadrian's Wall. Written on thin wooden tablets, these rich artifacts record official military documents along with personal matters.[11] Several of the 752 tablets reference beer, including requests for supplies because beer rations were low and record the name of Britain's first brewer: *Atrectus ceruesairus* (Atrectus the brewer).

The Americas remained isolated from the people, ideas, and cultural innovations of Europe, Asia, and Africa. Beer in this part of the world was invented independently and became fundamental to the social life during the Neolithic Revolution in Central and South America. Chicha, a beer made from corn, was intrinsically rooted in all aspects of social, economic, and spiritual life. Numerous artifacts indicate the importance and prevalence of beer to early societies. Archaeologist uncovered 7,000-year-old South American pottery that most likely stored chicha.[12] Scientists analyzed hair samples from 2,000-year-old mummies found in Peru and determined that their staple diet consisted of corn, seafood, and beer. The drink became central to the religious, social, and economic roles of the Inca Empire. Human sacrifices were rubbed in the dregs of chicha, and the mummies of kings were ritually bathed in maize flour and presented with chicha offerings. Throughout the Inca Empire, beer was the foundation to their economy; workers were paid in rations of chicha, which they consumed in great quantities during and after work.

Cerro Baul was an ancient political outpost and ceremonial center located in the mountains of southern Peru. For nearly 500 years, it was the religious center for the Wari Empire. Archaeologists have found the remains of

---

11. Pearce, J. 2002.
12. Hayashida, F. M. 2008.

an enormous ancient brewery, dating to about 700 to 1,000 AD.[13] The brewery contained a fire pit with evidence of animal dung apparently used to boil the water that created vast quantities of a spicy beer made from corn, used for ritual intoxication. Although there is no conclusive evidence as to the demise of the Wari people, most researchers agree that Cerro Baul was purposely evacuated based on the burning of roofs and smashing of vessels. However, before they picked up and left, evidence suggests they took over a week to brew a ceremonial beer and had an enormous party where they consumed it in vast quantities. Evidence suggests that the chicha was produced by high-status woman and presented first to the 28 local lords, followed by lesser lords, four senior leaders, and then the artisans, workers, and public citizens.

Patrick McGovern is a biomolecular archaeologist at the University of Pennsylvania Museum in Philadelphia. He has assisted brewers in recreating ancient recipes, most notably Dogfish Head Brewery's Midas Touch, inspired by the chemical analysis of vessels found in a 2,700-year-old tomb from today's northwest Turkey. The museum has a great collection of ancient artifacts and standing exhibits worthy of a visit, and McGovern, who has authored several books on ancient alcohol, periodically gives talks on the subject.

Dock Street
Brewing Company

Date of Visit: _____

Sampled _____

_____

Comments: _____

_____

The brewery closest to the museum also has historic importance. **Dock Street Brewing Company**, founded in 1985, was one of the nations' first post-Prohibition craft breweries and grew to be the nation's 26th largest brewery by 1996. Named for the historic Dock Street district of Philadelphia — the largest producer of beer in the county in the late 1700s — the company underwent transition in the early 2000s but relocated in 2007 to a beautifully renovated

---

13. Moseley, M. E., et al. 2005.

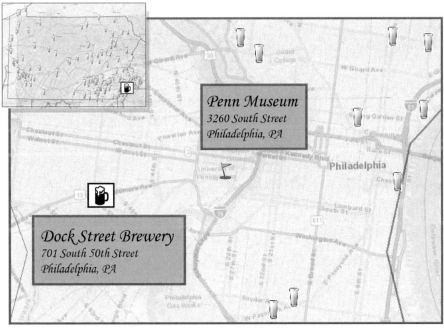

Penn Museum
3260 South Street
Philadelphia, PA

Dock Street Brewery
701 South 50th Street
Philadelphia, PA

110-year-old firehouse. The pub has a truly bohemian feel with an eclectic clientele that represents the great diversity of West Philadelphia. Along with a few trusted brews, they boldly venture into the unexpected with exotic recipes, including a one-time experiment with goat brain. Along with the great beer, customers can indulge in amazing stone-ground flour, hand-thrown, wood-fired pizzas.

Beer has not only been the focus of celebration; since the onset of civilizations, it has led to technological inventions, improved horticultural practices, and cultural innovations which, in turn, led to better nutrition and longer life expectancy. Is it a coincidence that the same time beer becomes a widely consumed dietary staple the world population doubles from seven to 14 million people?[14] Not that it was all positive. The Vikings managed to survive drinking their beer, which was often laced with ergot, a parasitic fungus that shares a similar chemical compound found LSD. Many Northern Europeans brewed with herbs such as mandrake, henbane, and mugwort, all of which can be used for painkillers, dream enhancers, and hallucinogens, with serious affects to the nervous system.

Despite some of the more powerful intoxicants added to beer, historically this fermented beverage was seen as a safer, healthier alternative to drinking water. Although not yet scientifically understood, people throughout history recognized that they did not get sick from drinking beer. Today, it is understood that the brewing process kills the microbes found in water, along with the added benefit that beer provides well-needed calories. Additionally, its warming, uplifting qualities were advantageous during a time when aspirin, anesthesia, cough medicine, pain killers, and tranquillizers were not available.

Beer remained a staple dietary fixture that was embedded in social customs throughout Europe as tribal societies organized into more formal territories and eventually into political states. Countries with different laws, regula-

---

14. Phillips, Rod. 2014.

tions, languages, and customs developed, and socially stratified populations emerged with differences in wealth, power, and education. Brewing
beer remained a domestic chore with a wide range of regional variations
and local recipes made from any available ingredients and local water.
However several places were beginning to create guilds, standardize methods, implement regulations, and transform brewing into a profession. With
this professionalism came the need to make a consistent product in larger
batches. Over the centuries, beer making centers emerged in parts of what
is today Germany, Luxemburg, and France.

During the sixth and seventh century, monasteries became educational
centers in Europe, recording literature and philosophy. The monks were
self-sufficient and took an oath to follow a devote lifestyle of poverty, obe-
dience, and contemplation. They dedicated numerous hours of physical
labor every day to their fields, bakeries, and breweries. The illustrious
monks created more bread and beer than they needed, and although they
lived a solitary life, they believed in hospitality and charity. Monasteries
became renowned places of refugee with good food and drink for travelers
and were particularly noted for brewing exceptional beer. The well-educated monks took a scientific approach to brewing. They experimented
with new techniques and ingredients and maintained detailed notes on the
results. In 1040, the Benedictine Abbey of Weilbenstephan, in Bavaria,
Germany, was granted the first official brewing license and the right to
sell beer for a profit, making it the oldest known continuously operating
brewery in the world. Increasingly, the word of good beer spread throughout Europe and the demand for monastic beer increased. By the 10th and
11th centuries, monks enjoyed a monopoly with increased wealth that
forced the secular lords and rising mercantile class to recognize beer's potential for profit.

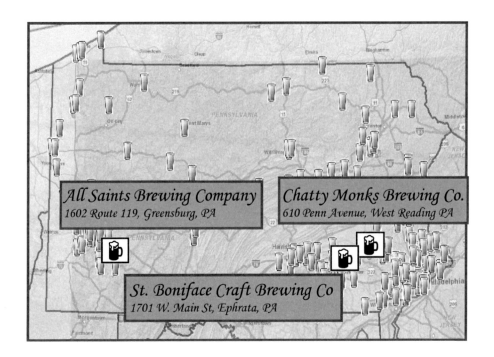

All Saints Brewing Company
1602 Route 119, Greensburg, PA

Chatty Monks Brewing Co.
610 Penn Avenue, West Reading PA

St. Boniface Craft Brewing Co
1701 W. Main St, Ephrata, PA

St. Boniface Craft Brewing Company

Date of Visit: _____

Sampled _____

_____

Comments: _____

_____

Monasteries of the past still inspire brewers today. Since beer was such an important part of life, many saints hold the title "patron saint of beer." A particularly noteworthy one is St. Boniface. Born in England, he is better known for his work converting Christians as he traveled around Europe. **St. Boniface Craft Brewing Company** opened in Ephrata in 2010 and honors the patron saint by welcoming travelers and introducing the masses to craft beer. The brewery has evolved and expanded over the years, to where they now have a 15-barrel brew house and a tap room with a brick oven pizza, long bar, open seating, and a delightful patio.

Boniface Wimmer, a Bavarian monk, immigrated to Latrobe in 1846. He helped spread the monastic brewing tradition by establishing the first Benedictine monastery in the United States, which later became St. Vincent, a small liberal arts college in 1870. St Vincent alumni, Jeff Guidos, opened **All Saints Brewing Company** in the nearby town of Greensburg. He uses his degree in chemistry to craft his "divine" beers including Heavenly Hefeweizen and Crimson Halo. Located in a large, renovated building that locals refer to as "the day old bread factory" **All Saints Brewing Company** has a large interior space, along with a great multistory deck that allows for large community events.

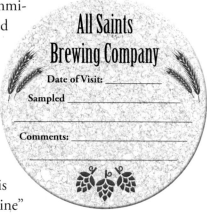

All Saints
Brewing Company

Date of Visit: _____

Sampled _____

_____

Comments: _____

_____

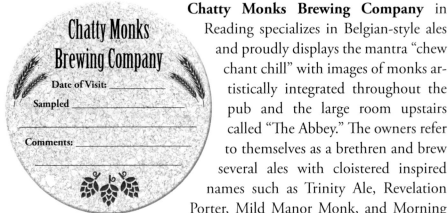

Chatty Monks Brewing Company in Reading specializes in Belgian-style ales and proudly displays the mantra "chew chant chill" with images of monks artistically integrated throughout the pub and the large room upstairs called "The Abbey." The owners refer to themselves as a brethren and brew several ales with cloistered inspired names such as Trinity Ale, Revelation Porter, Mild Manor Monk, and Morning Monk that pair well with the Hoppy French Friars and Forbidden Nachos. In the spirit of illustrious monks, **Chatty Monks Brewing Company** has resourcefully utilized materials from old barns, churches, and even debris to create a minimalistic, rustic feel. The wooden planks adorning the walls of the Abby and the beams used to frame the windows are recycled from an old barn that once served as part of the Underground Railroad. Continuing in the tradition of providing solace, **Chatty Monks Brewing Company** contributes to the revitalization of Reading and sits amongst a collection of unique stores and restaurants in a vibrant town.

As beer became a more professional, economic endeavor, the names, ingredients, and methods of brewing became standardized and regulated. For over a thousand years in Europe, ale referred to beverages created with malt, yeast, and water whereas beer included a range of available ingredients. By the 10th century, beer often contained honey, and by the 15th century, it contained hops. William IV, The Duke of Bavaria, imposed a standard in 1516, called the Reinheitsgebot, or German purity law, which stated all beer must be made from three ingredients: water, barley, and hops. This standardized the commercial profession and may be the oldest food-quality regulation still in use in today.

Gunpowder
Falls Brewing

Date of Visit: _____
Sampled _____
_____
Comments: _____
_____

Gunpowder Falls Brewing, located in Shewsbury Township, is the only brewery along the mid-Atlantic coast that is solely dedicated to brewing only lager beers and strictly upholds the Reinheitsgebot purity law. Head brewer Martin Virga obtained his brew master certificate from Doemens Brewing Academy in Munich, Germany and refined his craft in Germany. He returned to the United States, and after working at several other breweries, opened **Gunpowder Falls Brewing** in 2012. He proudly brews Pilsner, Export Hell, Dunkel, Helles, Schwarzbier, and Märzen, all in the style they did centuries ago with only four ingredients: water, yeast, hops, and malted barley. These authentic, German-style lagers can be appreciated in a tasting room adorned with German and Bavarian flags.

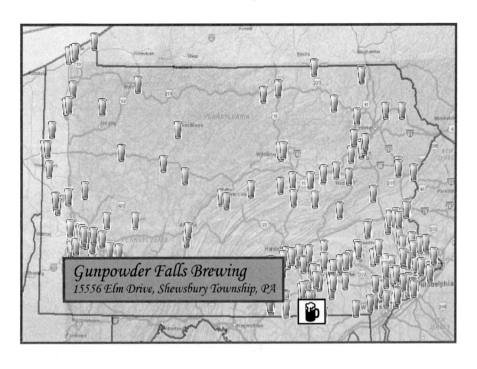

Gunpowder Falls Brewing
15556 Elm Drive, Shewsbury Township, PA

The agricultural science behind beer's main ingredient, hops, evolved as well. Hops add bitterness and aroma to beer and act as a preservative. Part of the cannabis family, this perennial grows in northern latitudes where cooler winters force dormancy. The wild plant was first documented by the Roman naturalist, Pliny the Elder, nearly 2,000 years ago, but it was not cultivated until 736 AD, and it was not documented in brewing until Abbot Adalhard in 822 AD recorded it in a recipe from a Benedictine monastery in Northern France. In the mid-1100s Germany and the Netherlands started to cultivate and use hops extensively as a preservative in their beer. It is believed to have arrived in England by the 1400s. For the next few centuries the distinction between beer and ale tethers on the presence or absence of hops. Many municipalities attempted to preserve the distinction of British ale by regulating and banning hops, but the northern climate created ideal growing conditions, and within 100 years hops were being home-grown in England rather than imported. By 1603, an Act of Parliament tried to maintain the quality of British hops to be the best in the world.[15]

15. Filmer, Richard. 1998.

Other European nations mandated the growth of hops, indicating how vitally important hops were to sustaining the health of family households and local farmers and to help regulate a country's economy. Hops is the only agricultural crop dictated by Swedish law. It required every farmer in the 15th century to have at least 50 hop poles. The regulation increased to 200 poles by the 17th century.[16] This stipulation limited imports from other countries and provided nutritional security. In a northern climate with unpredictable weather, hops preserved beer, which accounted for approximately one-third of the population's daily caloric energy needs, but it could also be used as means of saving germinated barley that would otherwise spoil as a food source. During poor growing seasons with limited food, the larger hops brines could feed animals and the smaller shoots, similar to asparagus, could feed people.

Hop Farm Brewing Company is located in Pittsburgh's revitalized Lawrenceville neighborhood and grows much of their own hops and other ingredients at their nearby farm. Owned and operated by Matthew and Emily Gouwens, their sustainable practices produce high quality hoppy and farmhouse ales, sours, and a collection of other creative beers. One of their recent recipes included 400 pounds of Michigan tart cherries that created an excellent "Cherry Bomb" sour which paired well with their farm-fresh menu items. Their signature "beer to burgers" meat comes from a local farm that feeds the cows and lambs **Hop Farm Brewing Company**'s spent grains.

Hop Farming
Brewing Company

Date of Visit: _____
Sampled _____
_____
Comments: _____
_____

---

16. Nilsson, Pia. 2001.

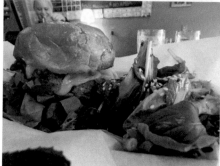

The prevalence of hops in Great Britain during the age of exploration, settlement, and colonization created one of today's most popular craft beers. India Pale Ales (IPAs) are original English beers that contain more hops than other types of beer, and they were created for the specific purpose of overcoming physical geography. Established in the 1600s, the East India Company had a monopoly between India and Britain and transported goods through varying temperatures, starting in the cold waters of England, crossing the equator, rounding the Cape of Good Hope and back to warm Indian waters. Exploding casks and spoilage was not uncommon during this long voyage. The amount of hops was increased in the beer as a preservative effect and to improve shelf life. The lighter color IPAs were more refreshing looking than the darker traditional ales, and with the repeal of the glass tax in 1847, which made clear glass more affordable, people could actually see what they were drinking. IPAs became the preferred drink of the higher classes and seen as a status symbol to those colonizing India and many other parts of the world.[17]

17. Haugland, Jake E. Vest. 2014.

Nimble Hill Winery and BrewingCompany

Date of Visit: _____

Sampled _____

_____

Comments: _____

_____

Settlers to North America transplanted hop genotypes, and the geography on the east coast was ideal for the plant to grow, both in cultivated fields and in the wild. **Nimble Hill Winery and Brewing Company,** located in Northeast Pennsylvania, produces most of their products from ingredients grown on their beautiful agricultural lands. They cultivate 1-acre of hops that grow alongside approximately 10-acres of vineyards. A cozy tasting room, located along US Route 6 in Tunkhannock, allows visitors to sample both wines and beers. Commonly available for tasting, bottle purchases, or growler fills is Hop Bottom IPA. This citrus IPA honors a nearby town, Hop Bottom, that was named for the plentiful and productive wild hops that grew along the local rivers and creeks.

Nimble Hill
Brewing Company
3971 US-6
Tunkhannock, PA

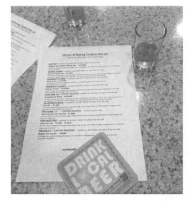

Hop farms were so important in Europe that it is possible they established some of the first land use practices and the initial economic foundations of the agrarian revolution. Increased productivity, crop rotation, fertilization, farming specialization, and commercial cash crops most likely emerged in the 1750s. The global outcome of these unprecedented inventions led to an increased food supply, which in turn increased world population, urban growth, and eventually the Industrial Revolution. However, studies of local cadastral maps from the 1600s indicate that hop specialization was well-established a hundred years prior to the agrarian revolution. While many smaller farmers had 50 to 100 poles of hops, adequate to sustain one-household, an emerging group of farmers had up to 8,000 poles. These medium-to-large scale hop farms had more manpower, extra resources, and a reserve of cash, which contributed to the development of an emerging middle class and challenged the centuries-old traditions of the nobility ruling a rural peasantry. The complexity of events that transpired in Europe cannot be oversimplified, but certainly the demand for beer and the need to grow hops impacted the agrarian, industrial, and political revolutions that advanced the world into the modern era.

Today, IPAs are often the signature beverage in many craft breweries. Brewers pride themselves on experimenting with creative ingredients, but all are founded on a critical element: hops. Hops are predominantly grown in the Pacific Northwest, with Washington accredited with 75 percent of the UnitedStates' yield, and Oregon and Idaho constituting the remainder of the total yield.[18] Historically, Pennsylvania boasted substantial hop agriculture, but downy mildew disease and Prohibition ceased production. A resurgence of local hop farms is emerging, and they contribute to the state's diverse agriculture and increasingly provide local ingredients to breweries. Many farms such as Sunny Brea Hops in Carlisle, Central Penn Hops in Halifax, Vista Farm in Orefield, Hop Hill Farm in Fleetwood, Hops on the Hill Farm in Easton, Barefoot Botanicals in Doylestown, and Bucktail Farm in Richfield are planting rhizomes. Most of these farms plant anywhere from 100 poles to 2 to 3 acres of hops, and pride themselves on growing and hand-picking top quality ingredients for craft breweries. Additionally, some breweries have purchased land or have erected a few poles near their establishments to grow their own hops.

---

18. Kopp, Peter A. 2014.

According to Beer Advocate, **Victory Brewing Company**'s Dirtworf Double IPA ranks as the top IPA in Pennsylvania. It is loaded with Citra, Chinook, Mosaic, and Simcoe hops that create an 8.7 percent ABV, full-bodied drink with a strong, intense aroma. Along with Dirtworf Double IPA, HopDevil, Helles Lager, and Prima Pils are a few of **Victory Brewing Company's** flagship brews that are well known for their hopped-up taste. They are one of the few breweries that regularly use whole flower hops in their beer rather than dry pellets. Although storing and preserving the whole flower is more difficult and it absorbs more wort, resulting in a decrease in volume, the essential oils and compounds in the whole flower creates a full- flavored, more aromatic beer.

Victory Brewing Company was founded in 1996 and is headquartered in Downingtown, but has opened additional pubs in Parkesburg and Kennett Square. Despite the growing success and increased production, the company has remained steadfast stewards to the environment. Their main production brewery is located in a renovated Pepperidge Farm factory. It houses a large restaurant and extended bar overseeing many brewing tanks and a small gift store. Along with the typical chalkboards found in most breweries, several LED boards promote the company's commitment to alternative energy and sustainable business practices. In their Parkesburg location they have a designated Victory Garden that supplies their chefs with fresh produce and give their employees space for their own personal plots. They regularly donate cases of beer to charity events and host events at their brewpubs supporting community organizations.

Prior to the Industrial Revolution, beer was made on a very small, local scale, but new inventions and the rapidly expanding global exchange of knowledge created everlasting change not only to the beer industry, but also to modern medicine, businesses, factories, and social regulations. Scientific instruments were invented that helped measure and regulate methods in the brewing process. Michael Combrune was a brewer in Hampstead, England. In 1758 he published the first writings on the recommendation and use of thermometers to manage the brewing process.[19] Nearly a hundred years later, the Bohemian chemist, Carl Joseph Napoleon Balling invented a hydrometer.[20] Along with other hydrometers invented at that time that measured the density of a liquid compared to the density of water, his also measured the amount of dissolved substances in the wort that was created during brewing. The hydrometer detected the exact amount of sugars, proteins, minerals, vitamins, and aromatics that help the brewer precisely monitor the alcohol levels in beer.

Beer was the impetus behind new inventions and allowed humans to gain a better understanding of their world.[21] Between 1840 and 1843, James Prescott Joule, a brewer in England, recognized the difference in heat used during the brewing process. He had access to apparatus and equipment and possessed chemistry skills learned during the beer making process, so he experimented with energy and heat. His published work led to the First

---

19. Combrune, M. 1804.
20. Meussdoerffer, F. G. 2009.
21. Shears, Andrew. 2014.

Law of Thermodynamics and his scientific achievement is recognized by his namesake Joule, used as the International standard unit of energy. James Harrison who is recognized as the inventor of refrigeration and who had recently created the first mechanical ice making machine was commissioned by a brewery in 1856 to build a machine that could cool beer. His inventions have had an immeasurable impact on the food industry. Just a few years later in the 1870s, the Franco-Prussian war cut off beer supplies to France from Germany. This lack of good beer made people question why beers originating in France and other countries often had a sour taste. Louis Pasteur was commissioned by the French government to study beer, and he concluded that the plagued flavors came from bad bacteria. He determined that beer was fermented by microorganisms, not chemicals, and with refined techniques of steaming beer slowly brewers could eliminate the bacteria. It's hard to imagine the modern food industry today without his widely employed technique, known as pasteurization, and more importantly, modern medicine without his contributions to germ theory and immunology. And anyone who has survived an introductory statistics course probably remembers the t-test, which determines if the average of two sample populations are statistically different. This test was introduced by a brewer who devised a cheap way to monitor the quality of stout beer while working for Guinness. Gosset published his work in 1908 under the pseudonym "Student" because Guinness' policy forbid their workers to publish their findings, hence the name Student's t-test.

Beer has had an extensive undeniable link to human history which still holds true in the modern world. The beer industry today still contributes to new inventions, changes cultural practices, and is deeply embedded in our social customs. Beer is economically important and builds regional and cultural ties. European nations such as Germany, England, and Ireland pride themselves on their beer culture and openly embrace their drinking production and consumption. Many places that Europeans colonized have inseparable historic and cultural ties to beer. Several key beer advertisements and logos, such as Molson's campaign "I am Canadian" and Foster's "It's How to Speak Australian" define entire countries by their connection to beer.

In the United States, places like Philadelphia, Pennsylvania; Milwaukee, Wisconsin; and Golden, Colorado have strong historic roots, whereas cities like Portland, Oregon and Seattle, Washington are thriving new hubs that overtly promote their connections to beer. The "beer test" has become a common phrase in American politics where voters consider who they would rather have a beer with to determine personality and likability of candidates. So in moderation, embrace that Liquid Lunch Club or repay a helping friend with a six-pack of beer, and take pride connecting with your earliest civilized ancestors. Remember, according to Gilgamesh, it is the LOVE OF BEER that makes us innately human.

CHAPTER 2

Beer Takes Over the New World

Beer and society have gone hand-in-hand for millenniums so it wasn't surprising that Europeans brought beer to their colonial settlements around the world as an essential dietary element for men, women, and children of all social ranks.[22] Despite the running streams that provided plenty of fresh water on these newly found lands, cultural biases against drinking water were deeply engrained. It was a utilitarian, cultural practice and a health habit.

Each European country had slight variations in objectives for colonizing the New World, and individual communities and settlers had personal economic, religious, and political reasons for leaving their homeland, but they all shared a few common goals: survive the journey across the Atlantic, construct shelter, find food, and build a brewery. Beer literally saved many explorers and settlers lives.

It's not surprising that William Bradford, the leader of the Plymouth party, prioritized beer in his list of dwindling provisions. Future ships like the *Arabella*, which brought Puritans to Boston, set sail with three times as much beer than water to survive the Atlantic crossing.[23] The British settle-

---

22. McWilliams, James E. 1998.
23. Lender, Mark Edward and James Kirby Martin. 1982.

ments of Jamestown, Raleigh, and Plymouth document building a brewery
as their first priority. In 1637, Captain Robert Sedgwick of the Massachusetts Bay Company was issued the first formal brewing license in the New
World.[24] French explorer Jacques Cartier, in Canada, would have died from
scurvy if his party had not brewed a spruce beer.[25] The Dutch settlers were
eager to explore the New World, with Henry Hudson traveling up rivers
looking for the Northwest Passage. Although they had better supplies than
the English in Virginia, their first priority was to establish a brew house
immediately upon arrival.

The Dutch settlers made several contributions with lasting impacts to
brewing in the New World. They began importing hops and developing a
commercial center for brewing in what is known today as Manhattan. In
1613, Adrian Block and Hans Christiansen converted a log house on the
southern tip of the island into a brewery. Historians recognize this establishment as the first commercial brewery in North America.[26] In the same
neighborhood, a few years later in 1645, The Red Lion Brewery started
doing business. This is the first identified brand of beer, creating an identity
with a new name rather than simply associating the beverage and the tavern with the name of the owner. Breweries in this neighborhood had a
lasting impact on urban development with another first in the Americas:
Brouwers Straat (Dutch for Brewers Street). This street was a hub of breweries and related businesses and was constantly wet as brewers dumped
their wastewater into the street. As more and more similar businesses were
established near one another and heavy-laden beer wagons transported
goods in rutted, muddy streets, the problem worsened. The solution was to
cover the road with stones. Brouwers Straat became known as Stone Street
and is the first paved street in America. It still exists under that name and
runs between Broad and Whitehall in New York City.

24. Cavicchi, Clare Lise. 1988
25. Ebert, Derrek. 2007.
26. Smith, Greg. 1998.

The original settlers and prominent Founding Fathers all brewed beer.[27] John Winthrop and his son John Winthrop, Jr., governors of Massachusetts and Connecticut, respectively, brewed a beer made out of native corn. Ben Franklin is often remembered for his delightful quotes, such as "Beer is living proof that God loves us and wants us to be happy," and "In wine there is wisdom, in beer there is freedom, and in water there is bacteria." Samuel Adams Sr. was known as a master beer brewer who passed the family business along to his son, Sam Adams, Jr. His cousin, John Adams, was also actively engaged in brewing. Thomas Jefferson had his own brewery at Monticello. George Washington recorded and published his brewing recipes, and in 1789, he declared his "Buy American" policy indicating that he would only drink American-brewed porters.

Tom Kehoe founded **Yards Brewing Company,** located in Philadelphia, in 1994. Along with their progressive initiatives — the first 100 percent wind powered brewery, using recycled materials to build the bar, and providing farmers with spent grain — they also reflect on their local history. **Yards Brewing Company** produces Ales of the Revolution with recipes from our Founding Fathers. General Washington's Tavern Porter comes from a recipe that is housed in the New York Public Library. It was part of a letter from General Washington to his field officers during the Revolutionary War. Tom Kehoe worked closely with the Philadelphia City Tavern and Monticello to recreate the Thomas Jefferson's Tavern Ale, and Poor Richard's Tavern Spruce is based on Benjamin Franklin's recipe. Visitors interested in sampling beers from the past can enjoy a Revolutionary Flight.

Yards Brewing Company

Date of Visit: _____

Sampled _____

Comments: _____

---

27. Lender, Mark Edward and James Kirby Martin. 1982.

Securing beer's importance in the New World can be attributed to the settlement of Pennsylvania with its diverse population living near abundant natural resources. Pennsylvanians changed the quality, quantity, and styles of beer forever. In 1681, William Penn acquired land from the King of England. Immediately, Penn understood the importance of securing friendly relations with the natives and exchanged commodities, including beer, for property rights. He established a colony with liberal provisions, religious tolerance, and economic opportunities and distributed pamphlets throughout Europe written in English, Dutch, French, and German. Incentives to relocate included land at bargain rates of a penny an acre or 200 acres for 100 British pounds.[28] This had an enormous appeal to many who had been excluded for centuries by the land-owning classes of Europe. Immigrants from all over Europe entered through the port of Philadelphia, which quickly blossomed into the largest town in the New World.

Hidden River Brewing Company is located on lands originally owned by William Penn, who sold it to one of the New World's earliest settlers. The original farmhouse structure still stands but is encompassed by a larger structure because the homestead had several additions over the centuries. This 300-year-old home, now referred to as Brinton Lodge, stands as a nonprofit historic site with daily tours and community events throughout the year. **Hidden River Brewing Company** probably has one of the more

28. Klett, Guy S. 1948.

offbeat claims to fame. The Briton Lodge is ranked first in the top 10 most haunted houses in Pennsylvania. The brewery teams up with Manatawny Creek Winery and Manatawny Still Works for the "Toast with a Ghost" tour. If non-paranormal sipping is more your thing, the brewery that is located in the historic lodge serves small-batch beers and a light menu sourced from many nearby farms. A portion of the proceeds contributes to the preservation and restoration of the lodge.

Hidden River Brewing Company

Date of Visit: _____

Sampled _____

_____

Comments: _____

_____

Several early settlers were noted brewers. Their business success made them prosperous, and in turn, they become well-established citizens in the growing colony. In 1684, the city of Philadelphia had nearly 500 inhabitants.[29] William Frampton, a merchant from New York City, was granted a license by William Penn to establish the first commercial brewery. He located and operated the brewery on the southwest corner of Front and Walnut Street, along Dock Creek in Philadelphia. Soon after, Joshua Carpenter — whose brother, Samuel Carpenter, built the first wharf in Philadelphia — established another brewery. By 1693, Joshua Carpenter was the second richest man in the city, second only to his brother. Similarly, Anthony Morris established a brewery on Front Street in 1687, which became a family business for generations, amassing great personal wealth. Morris went on to be a prominent citizen in Philadelphia, fulfilling his duties as a preacher, judge, and eventually mayor. Within a few years, five breweries operated within a few blocks of each other.

2nd Story Brewing Company

Date of Visit: _____

Sampled _____
_____

Comments: _____
_____

The emerging city in the late 1600s focused around the important Delaware River, connecting the port with small cobbled streets, alleys, and quaint buildings with ground-floor businesses and residences above. Philadelphia's Old City developed into a thriving retail, commercial, and residential center. Elfreth's Alley is the nation's oldest, continuously used residential street. Just a few blocks away, **2nd Story Brewing Company** creates a great ambience to complement a tour of Old City and Philadelphia's historical sites. As you walk along the cobbled streets, it feels like walking back in time. As the name suggests, their brewing system sits on the second story where exposed brick, dark wood, and steel beams are tastefully integrated into the historic décor. Openings above the main bar allow customers to view the tanks from both the first and second floor.

---

29. Wagner, Rich. 2012.

2nd Story Brewing Company offers unique food items that use farm fresh ingredients. Fritzie's Lager, a recipe from the original brew master's grandfather, is frequently used in the food preparation. A favorite item using Fritzie's Lager is Fritzie's Fondue Fries, which pairs well with the Beer Battered Fish and Chips. Customers can reflect on the Philadelphia's integral role as the nation transformed from a European colony to an emerging nation as they enjoy a Declaration IPA, Old City Kölsch, Old City Burning Rauchbier, or other rotating beers.

Breweries were the nerve center of the colony, helping to secure Penn's lands as the nucleus for economic, political, and social activities in the New World. By 1700, Philadelphia surpassed New York and was challenging Boston as the cultural hub of North America. Beyond the city limits, the great agricultural lands of Penn's colony were also filling with diverse, illustrious immigrants. As the cities in the east became well-established communities, the newest arriving immigrants were typically in search of cheap, available land, and thus were forced to move west.[30] Outside influences such as cold winters, bad harvests, and famine brought waves of settlers from Europe, particularly Germans, English, Swiss, Dutch, and Scots-Irish.

Settlers were advised to bring barley, hop roots, and copper kettles with them to the New World. Imports were unpredictable and infrequent, so settlers had to rely on local production. Brewing beer was part of a woman's

---

30. Ross, Michael and Paul Marr. 2008.

domestic duties, and most kitchens were designed and built around caul-
drons that brewed beer. Beer was produced often and in small batches,
usually enough to supply a family for several days.[31] Estate records indicate
that the majority of settlers brewed their own beer and brewing equipment
was often listed as notable possessions bequeathed to daughters and wid-
ows in wills.[32]

Typically, beer was dark and cloudy, flavored with hops, and contained
approximately 6 percent alcohol.[33] However, in the rural countryside,
methods of brewing were rudimentary, supplies were commonly limited or
non-existent, and growing barley was unpredictable, so settlers creatively
used alternative ingredients like pumpkins, parsnips, seeds, berries, spruce,
birch, and sassafras. This unpredictable assurance of crops and food supplies must have been stressful to a self-sufficient farmer, but it created resil-
ient settlers with a resounding ability to improvise and concoct nutritional
substitutions. This lasting versatile trait, paying respect to the unpredictable nature of the seasons that transcends time, has created a contemporary
craft brewery favorite: the farmhouse ale. Originally from the French-speak-
ing region of southern Belgium, farmhouse ales or saison beers (French for
season) were created in the fall and stored for several months.

Tired Hands Brewing Company, located
in Ardmore, was inspired by the early
farmhouse brewers of America and
creates several award-winning saisons.
SaisonHands is their most accurate
interpretation of a traditional saison,
brewed with rye, wheat, and oats.
They also have notable saisons such
as Shambolic, Out of Emptiness, and
Sticky Drippy Crystals. In their original
location, **Tired Hands Brew Pub** serves a

31. McWilliams, James E. 1998.
32. Moss, Kay and Kathryn Hoffman. 2001
33. Smith, Greg. 1998

food menu that complements the farmhouse traditions with meats, cheeses, and baked-on-premise breads in a simple but eclectic style bar with views of the small brewing tanks. Less than a half-mile away, they opened **Tired Hands Fermentaria**, situated in an 85-year-old building that once housed a trolley repair shop. Here they can serve more customers with a larger seating capacity and a greater production facility. Additionally, they have a general store that sells bottles, growlers, and merchandise. Not only is **Tired Hands Brewing Company** committed to crafting award winning beers, they are dedicated to assimilating art throughout their buildings. Extending the theme of their wavy hands logo, interesting graphic designs form the central theme of their bike racks, t-shirts, framed pictures, stickers, and an elaborate wall mural by Mike Lawrence drawn with a sharpie — yes, a sharpie!

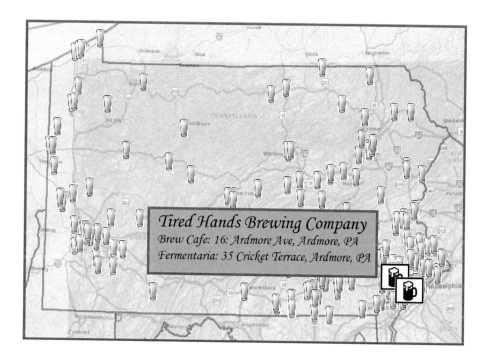

Tired Hands Brewing Company
Brew Cafe: 16: Ardmore Ave, Ardmore, PA
Fermentaria: 35 Cricket Terrace, Ardmore, PA

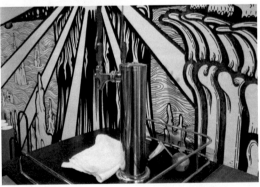

Commercial breweries were found throughout the colonies everywhere Europeans settled because of the storage and distribution technology of the time. Storing beer for an extended period of time was not possible for a few more centuries and transportation systems were non-existent. Unlike wine and spirits, the freshness of beer decreases relatively quickly, thus the highest quality and lowest costs are found closest to local production. Early breweries demonstrate a classic example of market-oriented manufacturing, locating production close to the point of sale.[34] Because beer was expensive to transport and had the potential to spoil, early breweries were limited to producing small volumes and reaching only local markets.

The quintessential tavern where beer was dispensed dominated the cultural life of early settlers. It was paramount to the community as the central location for the exchange of goods, services, and ideas. Travelers used them as hotels, which in turn increased business, brought in external money, and provided a communal link and network to other communities. Not surprisingly, tavern owners often ranked amongst the wealthier citizens in the community. Between 1750 and 1840, as the road networks became established, thousands of tavern licenses were petitioned and taverns existed on all well-traveled routes.[35] One great example is the Widow Piper's Tavern, in Shippensburg, built around 1735. Shippensburg is the oldest community in the Cumberland Valley and second oldest west of the Susquehanna River. This beautiful stone building was probably one of the first public taverns west of the Susquehanna and was vital to the economic and social growth of colonial society. Operated by Janette Piper, a recent immigrant woman whose husband died during their Atlantic crossing, the multipurpose tavern was an inn that welcomed weary travelers, provided news and information as a key stop along the postal route, and served as the first Cumberland County Courthouse from 1750-1751 before it was later moved to Carlisle. The building no longer serves as a tavern, but tours are available on the second Saturday of each month.

---

34. Shears, Andrew. 2014.
35. Smith, Greg. 1998

If you like the idea of enjoying a beer after a long day of travel in a historic inn, consider visiting the **Black Forest Brewing Company** in Ephrata. It sits on the beautiful property of the 1777 Americana Inn Bed and Breakfast — which was nationally recognized as one of top six bed and breakfasts for beer lovers across the United States by BedandBreakfast.com — and is located in Ephrata's commercial historic district. The neighborhood has over 30 historic buildings including a bank, post office, railroad station, and hardware store. The gardens of the inn blend into the patio of the quaint brewery, nestled amongst historic streets and important buildings where visitors can enjoy an idyllic, peaceful setting. Named for a hiking trail that inspired the owner and his three sons to embark on a new type of journey, **Black Forest Brew-**

Black Forest Brewing Company

Date of Visit: _____

Sampled _____

_____

Comments: _____

_____

ing Company maintains the historic theme with a colonial decor and serves six of their beers on tap, Pennsylvania wines, and mixed drinks from a local distillery.

The importance of these taverns to the social, economic, and political institutions of North America cannot be over stated. Taverns in colonial America forged the infrastructure of the nation. Recognized as essential meetinghouses for trade, marketing, and social functions, they were well-established structures. Townspeople and travelers would gather to drink beer, exchange news, and conduct business in a safe location. Political parties and campaign headquarters all started in taverns, and the foundations of democracy were argued and debated in taverns while people gathered to drink beer.

In a limited monetary society, beer was a critical commodity. It was often used, traded, and given as a means of payment. Communal projects, such as clearing a common field, raising a barn, or building a church proceeded with provisions of beer. In the recruitment of an army for a new country with limited funds, beer was essential. Not only was it given out as daily rations but it served as a socializing technique for neighbors during "drill days" and as an important commodity during recruitment. Unlike the comforts and security in their European homelands, every new settler was concerned with the defense of their family, home, and farm. Colonists commonly formed militias and volunteered for part-time armies to combat threats and were recruited, trained, and repaid in beer.[36]

---

36. Smith, Greg. 1998

Beer was considered healthy, nutritional, and fundamental to daily life. Because of this, it was commonly provided in numerous settings. Drinking beer was considered preventative medicine. Doctors often prescribed dark, rich porters for nursing mothers, the elderly, and to those who were sick or frail. Commonly, people would start their day with a breakfast beer and continue drinking at work. Beer was considered part of the benefits employers would provide to their workers. Before coming to the New World, John Harvard studied brewing from William Shakespeare. When he founded the university that bears his name, he made building a brewery a top priority to supply beer to both faculty and students. Similarly, when William and Mary University burnt down in the early 1700s, the brewery was the first building to be rebuilt. Later, Matthew Vassar emigrated with his family as a young child from England. Following in a long tradition of family brewers, he became very successful brewing ale along the Hudson River. The wealth generated from his brewery allowed him to engage in philanthropic activities. Committed to the idea that woman should be offered the same educational rights as men, he commissioned an architect and donated funds to build a female college that would compare to Harvard and Yale, hence the highly acclaimed Vasser College.

Levity Brewing Company

Date of Visit: _____

Sampled _____

_____

Comments: _____

_____

Most college towns in Pennsylvania today have a craft brewery nearby. Indiana University of Pennsylvania, one of the largest of the 14 schools in the State System of Higher Education, is no exception; in fact, they have two. **Levity Brewing Company** opened as Indiana, Pennsylvania's first brewery in over 75 years. In a renovated window manufacturing plant, the owners have retained portions of the old building — like the truck-dented ceilings and old delivery doors — but renovated with a community-centric and rustic theme: local timber flanks the walls, bars, and tables; oak barrels are used for extra storage and tables; artwork by local artisans is on display; and used grain bags from local maltsters serve as creative window shades. Many of the beer names reflect people's commonality and connections with education, such as "Summer School" and "Gymclass Allstar." A must try is "Tenured," a double IPA with 8.9 ABV. It's hopped but well-balanced and is dedicated to the distinguished spot that IPAs hold in craft brewing realm. Similar to many tenured faculty, the beer is described as somewhat crazy, somewhat dry, sometimes bitter, but always brilliant.

Levity Brewing Company taps 17 or 18 great beers at any given time. The additional beer names reflect their growing reputation for hosting great live music, including "Haze Frehley" and "B-side" (one of the best sours I've had in a while), and the local geography, such as "Indiana Autumn" and "Hoodlebug Brown." They pride themselves on their quality food items that are often infused with their beer and sourced from local farms.

Less than a mile away, this university town recently saw the opening of **Nobel Stein Brewing Company**. Started as a dream in Alex's garage with his brother Max and two of their friends, the name reflects the integrity of

Nobel Stein Brewing Company

Date of Visit: _____

Sampled _____

_____

Comments: _____

_____

their product and the long-standing tradition that beer is a vital element to humans. The hard work and determination of these brewers is evident in the quality of craftsmanship of their brewery. From the copper ceilings to the quarter-keg disguised sound system and the hand-crafted bar, **Nobel Stein Brewing Company** wants to embrace the community and provide an exceptional cultural experience that extends beyond quality beers.

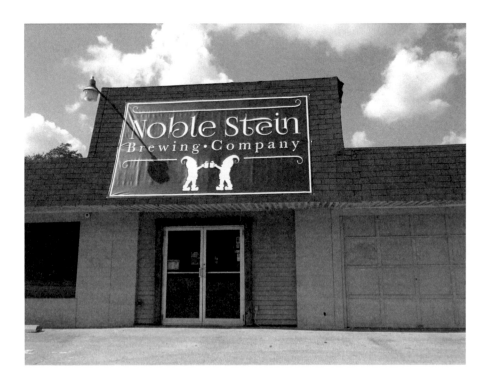

Immigration to the New World continued and Europeans moved west past Pennsylvania's borders. As people migrated and settled the west, beer accompanied them. Westward expansion, transportation, and the mining industry all account for the diffusion pattern of beer. A notable example is the number of breweries along the Erie Canal that followed its opening in 1825.[37] Not only did it serve as a major transportation route for people and commerce, the canal increased the production, distribution, and consumption of beer. By the mid-1880s, Wisconsin — a state known for its German immigrants — had over 300 breweries, with one in almost every community.[38] Farther west in the prairies, The Hudson Bay Company established a commercial brewery in 1847. By the time Winnipeg was settled in 1874, it already had seven breweries for a population of 1,869 people.[39] Breweries appeared in the Denver area by the 1860s. One of the most notable is the Coors brewery, founded by the merger of German immigrants

37. Batzli, Samuel. 2014
38. Shears, Andrew. 2014.
39. Ebert, Derrek. 2007.

Adolph Coors and Jacob Schueler in 1873. It has resiliently endured changing times, been incorporated into other businesses, and emerged as one of the largest breweries in the world today.

Pig Iron Brewing Company

Date of Visit: _____

Sampled _____

_____

Comments: _____

Abundant natural resources were one of the main incentives for westward migration, and Pennsylvania had its fair share. Numerous iron furnaces, along with canals and railroads, helped establish the first industrial complexes in the United States and helped make Pennsylvania a leader in iron production. Anthracite coal, local iron ores, and limestone produced pig iron that was easily transported and later converted into steel, cast iron, or wrought iron. Marietta is a historically important town that developed along the Susquehanna River. Visitors can walk or bike along the Northwest Lancaster County River Trail and enjoy the many well-preserved charming landmarks of an earlier industrial age. **Pig Iron Brewing Company** is located along the Front Street of Marritta, adjacent to the railroad tracks and the river. The inside bar and seating area have exposed brick and historic photos on the wall and the beautiful patio outside lets visitors enjoy freshly brewed beer, hard lemonade, and hard cream soda and soak in the historic scenery.

The 1849 California Gold Rush brought hard working immigrants all the way to the west coast, and they wanted beer. By 1852, San Francisco had over 350 establishments selling the highly demanded beverage.[40] In 1871, German brewer Gottlieb Brekle bought an old saloon and started Anchor Brewery. The company survived Prohibition and fires, along with changing management, for nearly a century. It eventually closed for a few years but was revived in 1965 by Fritz Maytag, who is often cited today as the inspirational father of the craft brewery revolution.[41] Just farther north, Henry

---

40. Erickson, Jack. 1987
41. Flack, Wes. 1997.

Weinhard, a German immigrant, moved to Portland, Oregon in the 1850s and took over an existing brewery. It too survived Prohibition with varying success in the 1960s and 1970s, and today the brewery is an architectural fixture in the trendy industrial neighborhood known as the Pearl District. Although the company rebranded itself as a staple in the craft brewery movement of the 1980s, it yielded to the pressures of big business and was incorporated by SAB Miller in the 1990s.

Westward expansion, settlement of the continent, and growth of a nation continued, and the beer followed. Breweries were thriving across the entire continent. In 1873, the industry peaked with a total of 4,131 breweries throughout the United States.[42] The late 1880s was a time of rapid inventions and technological advancements throughout the United States. The emergence of mass-produced beer in the late 1800s and early 1900s can be accredited to the development of transportation networks and improvement of bottling technology. Chicago developed as a major Midwest City with a thriving population due to its agricultural and manufacturing ties. The Great Fire of 1881 destroyed five of the 12 large breweries in the city.[43] The destruction led to an opening in the market for neighboring breweries in Milwaukee. The success of several companies in this nearby city provided the extra capital needed to invest in new mechanization and changing technology.

One of Pennsylvania's largest contribution to the movement of people and goods to the western states is the railroad, and one of the greatest engineer feats of the railroad industry was building the Horseshoe Curve in Altoona. It was built in 1854 to lessen the extreme grade of the Allegheny Mountains. Altoona is known as the Railroad City, and the horseshoe curve has remained a vital component to the industrial heartland. It was such a crucial element to moving natural resources, goods, and people that during World War II, Nazi Germany sent saboteurs to destroy the curve — they failed. **Railroad City Brewing Company** occupies a visually impressive building in downtown Altoona and creatively uses railroad signs, spikes as

42. Flack, Wes. 1997.
43. Shears, Andrew. 2014.

tap handles, and old doors to help visitors appreciate the industrial spirit of the railroad.

Industrialization accounts for much of the growth in the production of beer in the United States over the past century, and family businesses like Miller, Busch, and Pabst established strong-holds in the market. The big compa-nies adopted pasteurization by 1878, enabling them to ship kegs of beer globally, and ice machines were intro-duced at the Centennial International Exhibition in 1876. Hosted in Philadel-phia, it was the first official World's Fair held in the United States and displayed a range of new inventions that revolution-ized mass production. Companies that could afford to invest in technology had a tremendous advantage over their smaller competitors.

Railroad City Brewing Company

Date of Visit: _____

Sampled _____

_____

Comments: _____

_____

Pabst became the largest brewery in America, marketing to every state, and Anheuser-Busch eliminated the middleman by building their own taverns that sold only their beer along the new network of rail lines. Smaller brew-eries could not compete and eventually went out of business while the larger companies merged. The decline in the number of breweries contin-ued into the next century when the Temperance Movement — spurred on by religious and political change, the declaration of war on Germany, and the passing of Prohibition in 1917 — led to the demise of most local breweries. Pittsburgh, for example, had 36 breweries in 1900 and only two by 1910.[44]

Following Prohibition, the United States' beer industry experienced a steady economic growth, along with the consolidation of breweries, to enter an era of mass production. The largest five firms grew in sales from 10 percent in 1910 to 24 percent in 1950, to 75 percent in 1975, to over 80 percent in 1990.[45] Anheuser-Busch, Pabst, Miller, Blatz, and Schlitz-Stroh held the top tier of the market. The consolidated businesses experienced economic restructuring. Prior to World War I, beer was primarily sold in taverns, stored in kegs, and occasionally sold in bottles. Canned beer debuted in 1935, providing a cheaper alternative, and by 1942, nearly 60 percent of American beer was packaged.[46] This changed consumption patterns. Rather than appealing to only men sitting in saloons or taverns, advertisements began showing people with friends and families enjoying beer at home, at picnics, and playing sports.

Sly Fox
Brewing Company
Date of Visit: _____
Sampled _____
_____
Comments: _____
_____

Today, many craft breweries are canning their beer for environmental, economic, and user-friendly reasons. The **Sly Fox Brewing Company** was one of the first craft breweries to start canning and has been an advocate of aluminum cans since 2006. Not only do cans get colder faster and chill beer better than glass bottles, the aluminum protects beer from UV rays, which can often spoil and skunk beer in bottles. Cans are a more compact shape, stack more easily, and require less packaging material than bottles. Addi-

---

44. Shears, Andrew. 2014.
45. Caroll, Glenn R. and Anand Swaminatham. 2000.
46. Erickson, Jack. 1987.

tionally, they weigh less than glass, so the transportation costs are significantly less. Cans are ideal for outdoor activities and are usually allowed in many more places where glass bottles are prohibited from fear of shattering.

Sly Fox Brewing Company opened in 1995 in Phoenixville. Although they now are located across the street from their original location, the Phoenixville location has indoor and outdoor seating and plenty of parking. With the growth of the company over the years, the Giannopoulos family and head brewer Brian O'Reilly have expanded **Sly Fox Brewing Company's** production and opened a second location in Pottstown in 2012. The large factory, located in an industrial park-like setting, allows for ample indoor and outdoor space. An expansive patio is great for friends, families, and leashed pets to relax, enjoy a beer and good food, and play a round of Frisbee golf. Glass walls allow visitors to see the impressive canning system and the amassed storage of packaged beer that they ship to five states and Washington D.C. It's not hard to understand why the company is recognized as one of the outstanding breweries in the Mid-Atlantic. In 2007, the Great American Beer Festival awarded Sly Fox Pikeland Pilsner a Gold Medal, the first medal of any sort to be given to a canned beer. Success and growth continued, and they now produce over half a million cans of beer a year and are experimenting with new designs. **Sly Fox Brewing Company** is the first craft brewery in the United States to debut 360 Lid, which have a wider opening than traditional cans. Smell is the dominant sense that affects flavor perception, so these topless cans, not yet legal in all states, allow for more sophisticated tasting of full-flavored craft beers.

In the mid-1900s, merging beer companies benefited from these changing consumption patterns and continued to increase their production. They created new styles and reduced production costs by adding corn and rice, which made the beers lighter in flavor and lower in calories. In the 1960s, Americans were starting to become health conscious, and Dr. Owades, a biochemist, made Gablinger's diet beer by removing starches. With limited marketing, this diet beer did not go over well with male consumers.[47] Miller acquired the company in the early 1970s, created Miller Lite, and began ad campaigns with macho celebrities arguing over "tastes great" or "less filling." Within the first few years, the diet phenomenon produced annual sales of 10 million barrels a year and reached as high as 19 million barrels by 1990. Labatt's, in Canada, launched an aggressive and successful mass marketing campaign that focused on lifestyle rather than the beer, stressing the 'image' of the beer rather than the taste. The other major companies soon copied with similar products.

Beer companies became enormous business conglomerates in the second half of the twentieth century. As the industry began to plateau in the 1970s, competition between these top four or five breweries increased and marketing and advertising became essential. Television and radio advertising expenditures increased fivefold from 1977 to 1998 with brewers spending $752 million a year. Anheuser-Busch, Miller Brewing Company, and Coors Brewing Company accounted for 85 percent of the advertising dollars.[48] This large monetary investment had a significant impact on market share and creative commercials played a critical impact in consumer behavior, particularly with a strong link between viewing sports and drinking beer. In 2011, in the United States alone, $10.9 billion was spent on advertising during televised sporting events.[49]

Beer, and its corporate sponsorship, became deeply engrained in the fabric of entertainment in the United States. Many fans find commercial messages entertaining rather than obtrusive, and nowhere is this more apparent

47. Kell, John. 2015.
48. Wilcox, Gary B. 2001.
49. Levin, Aron, et al. 2013.

than in one of the most widely televised sporting event in the United States: the Super Bowl.[50] Commercials have become hyped events rather than interruptions to the game. In the 1990s, Anheuser-Busch invested in exclusivity during the Super Bowl and developed memorable commercials, jingles, and beer products that became household names. Since then, years of Bud-bowl campaigns, the Bud-Wise-Zer frogs, and a lost puppy befriending a Clydesdale have captivated audiences enough that many viewers are more interested in the ads than in the outcome of the game.[51]

50. Yelker, Rama, et al. 2013.
51. McAllister, Mathews P. 2001.

Beer has sustained Americans since the first explorers left Europe. The development, growth, and success of the United States' economic, political, and cultural influences have been amalgamated with beer. In 2007 and 2008, the large American beer companies entered into multi-international corporations.[52] Common brands brewed by Budweiser, Miller, and Coors, are now part of AB InBev and SABMiller, merging with global companies, creating the largest breweries in the world, and recording tens of billions of dollars a year in profit. From humble beginnings and the hard work of many resolute immigrants, many breweries in the United States are a testament to the American dream. These family owned businesses developed into large, successful companies, many of which are iconic symbols sharing their LOVE OF BEER around the world.

---

52. Hoalst-Pullen, Nancy, et al. 2014.

CHAPTER 3

Pennsylvania Dominates the Beer Industry

Did you hear the one about the Englishman, German, and Irishman who walk into a bar? Lucky for us, the punchline is a state rich with exceptional beer flavors that reflect the diversity of the early European settlers. Pennsylvania not only embraced early settlers and their drinking traditions, it fostered them. The physical resources and cultural traditions set the stage for technological advances in the beer industry. Many firsts in America occurred because of beer in Pennsylvania, and those innovations had lasting impacts on the economy and culture of the state and the nation.

One of Pennsylvania's most creative and innovative citizens is well known for his love of beer. Benjamin Franklin was a leading author, scientist, inventor, civic activist, statesman, and diplomat. He was a postmaster and facilitated the development of Philadelphia's fire department. **Saint Benjamin Brewing Company** pays tribute to this Founding Father and is located in the Theodore Finkenaur Brewery, which in the 1880s produced 15,000 barrels of lager beer annually. Today **Saint Benjamin Brewing Company** is slowly increasing their production and have expanded to a

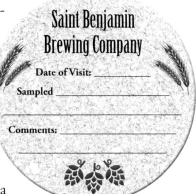

Saint Benjamin
Brewing Company

Date of Visit: _____

Sampled _____

_____

Comments: _____

_____

full tap room with open seating and views of exposed brick and large windows where customers can enjoy 12 brews on tap and a full food menu.

Originally in the New World, beer styles followed the British tradition of top-fermenting ales, porters, and stouts. Following the Declaration of Independence and the initiative to purchase American-made products, the production of porters doubled in Philadelphia. Although much of the production was for local consumption, Philadelphia brews were well known throughout the Atlantic. In 1793, the United States Commissioner of Internal Revenue reported that Philadelphia shipped more beer than the rest of the nation's seaports combined.[53]

Beginning in the 1820s, an increased number of Dutch, Irish, and Germans immigrants entered the port of Philadelphia, bringing with them new brewing styles and expertise. Most notably was the large influx of

---

53. Wagner, Rich. 2012.

Germans whose homelands had different ingredients, technology, and innovative ideas. Beer, which was already common in daily life, increased rapidly in popularity. Nowhere was this more evident than in Philadelphia. The city developed into a beer-producing hub. The abundant fresh water, fertile soils, moderate climate, and the city's location at the mouth of a large river made it a center for brewing. From 1612 until 1840 Philadelphia led the country with over 101 breweries in one city alone, nearly twice as many as the second leading city, New York.[54]

Several breweries led the nation in architectural achievements, economic growth, and technological advances, while notable personalities left a colorful history that had a lasting impact on the young, developing nation. Robert Hare, son of a brewer in London, immigrated to Pennsylvania in 1773 and started his own business. It is often claimed that he introduced North America to porters.[55] George Washington was a faithful customer, and Hare became a leading brewer in Philadelphia. He later joined forces with a German immigrant, merging their businesses into The Gual Brewery. After the American Revolution, Hare was elected to many prominent positions; he attended the state Constitutional Convention of 1789 and was a state senator, where he was later named Speaker of the Senate.

New inventions and industrial technologies were often first used by breweries in Pennsylvania. In 1819, The Francis Perot Brewery, which had acquired Philadelphia's first brewery established by Anthony Morris in 1697, installed one of the first steam engines used in America.[56] The machine ran for over 60 years, producing hundreds of barrels of beer a day. Of course, the success of the steam engine is well-noted in developed industrialized nations, and The Perot Company is widely recognized as the oldest, continuously operating business in America until it was sold in 1963.

One of Pennsylvania's greatest contribution to the beer industry was the introduction of lager beer. In 1840, John Wagner immigrated to Philadel-

---

54. Batzli, Samuel. 2014.
55. Smith, Greg. 1998.
56. Wagner, Rich. 2012.

phia from Bavaria with a different type of yeast that produced lager beer. A historic marker stands at 455 St. John Street, Philadelphia commemorating the location of the nation's first lager beer. Charles Wolf, a sugar refiner, purchased the yeast and in conjunction with another immigrant created the Engel and Wolf Brewery. They started to produce large batches of lager beer, which sold in great quantities to the growing German population. This bottom-fermenting beer grew in popularity, particularly with the working classes. Lagers are generally lighter in flavor, higher in carbonation, and golden in appearance, which Europeans found more appealing than darker, heavier ales in the hotter climate of North America. Consumption of lager beer increased enormously, and by 1865, lagers out produced and were more widely available than ales.[57] The new style of beer opened more opportunities for incoming Germans to build more breweries and, in turn, establish large companies and accumulate great wealth.

Louis Bergdoll emigrated from Germany to Philadelphia and established a Brewing Company in 1849. During its peak operation, the brewery created a cold lager beer that became one of the most popular beers in the country, with the brewery producing thousands of barrels a day. The company finally dissolved in 1951 through a series of declining events including Bergdoll shooting himself in the head, Prohibition, and the final sale of the Bergdoll building at 29th and Parrish Streets. However, by the end, the Bergdoll estates had amassed great wealth valued at several million dollars, and the brewery itself had transformed Philadelphia. The brewery required enormous amounts of water and ice, wood for barrels, and transportation of raw materials. The demands produced growth in related businesses, including refrigeration, air conditioning, and construction. The Bergdoll Brewing Company is responsible for transforming a neighborhood into a bustling industrial center known as Brewerytown.

Brewerytown flourished with breweries, keg factories, bottle manufacturing, and living spaces for workers. Moving into the modern era, the end of the 19th century saw the transition from the previously used method of cooling beer in caves to the newly invented electric refrigeration. This tech-

---

57. Batzli, Samuel. 2014.

nology was passed from the breweries to the local population with the establishment of public utilities provided by The West End Company, later the Philadelphia Electric Company. Germanic brick architecture is evident throughout the neighborhood in dozens of buildings ranging from large to small breweries, related storage buildings, stables, owner's mansions, and worker's row houses. Architect Otto Wolf, son of brewer Charles Wolf, is accredited with designing over 60 buildings in the neighborhood, while his rival, William Decker, focused on high profile buildings and mansions.

Together the architectural styles provide a sense of continuity in the neighborhood. Today, Brewerytown is recognized as a Historic District listed on the National Register of Historic Places. **Crime + Punishment Brewing Company** is proud to bring the tradition of brewing back to Brewerytown and be a part of the growing, thriving neighborhood and community. Many of the beer names, such as Chagall Window and Leon Trotsky Trout, and food items connect the local history with themes and characters in Russian literature

Crime + Punishment
Brewing Company

Date of Visit: _____

Sampled _____

_____

Comments: _____

_____

Other notable breweries filled the Philadelphia landscape to make great tasting, internationally renowned beer. Schmidt was another German immigrant who moved to the city in 1860 and acquired Courtenay's Brewing Company. He renamed the company using his last name and transformed it into a successful business. The company continued to grow, reporting some of the highest volume of sales across the entire United States. The Schmidt Brewing Company's sales ranked in the top 20 by the 1920s and peaked at 13 in 1959.[58]

Philadelphia thrived, and beer was one of the city's leading industries. By the 1880s beer was the fifth most valuable product after cloth, carpets, steel, and construction and generated more income than specialized industries such as railroad and engines. For a few decades after the American Revolution, beer production declined slightly. Beer was seen as British, and barley at times was difficult to grow. Whiskey gained in popularity, and the flourishing apple trees provided plenty of material for cider, but Americans still loved their beer. When the United States conducted its first census in 1810, 132 registered breweries — 98 of which were located in Pennsylvania — produced 185,000 barrels of beer.[59] The prevalence of beer increased rapidly. In only 40 years, the 1850's census recorded 421 breweries. Between 1841 and 1865, Philadelphia had over 190 breweries, whereas the second leading city, New York, had only 95. This trend continued through

---

58. Wagner, Rich. 2012.
59. Smith, Greg. 1998.

the next era of increased mechanization, 1866 to 1920, which witnessed Philadelphia leading the nation with 299 breweries, followed by New York City with 145.[60]

The success of the German style lager radiated from Philadelphia to surrounding cities, across Pennsylvania, and eventually throughout America. Fredrick Lauer emigrated from Germany, and established a brewery in Reading, which he later passed on to his two sons. The production of beer was quite lucrative, and the brewery became the third largest beer producer in Pennsylvania, bringing prestige to the entire family. The father turned down a nomination to run for Congress, but did equip the entire Pennsylvania 104th Company during the Civil War at his own expense. He was also elected and assumed the position of the first president of The United States Brewers Association, where Lauer established standards for the industry, promoted sound labor practices, and helped to establish fair and reasonable taxation.

Oakbrook Brewing Company continues on the path of brewing classic style ales in Reading. Located in a restored 1905 fire station, the tanks and bar are situated in the former horse stables, and the brewery is situated in the former engine house. Using historic photographs, the building was restored in 2016 to its early 1900s appearance, matching the décor, paint, fixtures, and floor plan as best documented. Photographs are displayed around the room to preserve the communities' history, and long beer-hall style tables encourage conversations with neighbors.

60. Batzli, Samuel. 2014.

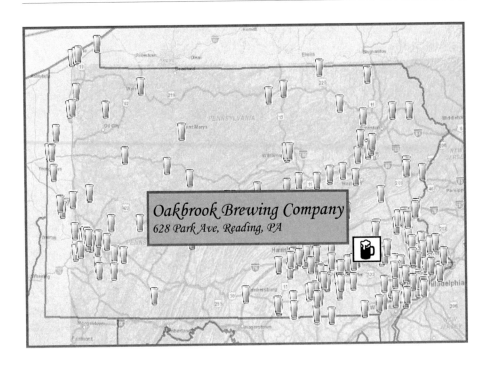

Oakbrook Brewing Company
628 Park Ave, Reading, PA

In 1876, Alois Bube, a German immigrant who learned his trade as a brewing apprentice in his homeland, bought a small brewery from Philip Frank in Mounty Joy. The demand for lager beer helped Bube grow a successful business; he expanded his brewery several times and built a Victorian style hotel to accommodate overnight guests. Workers at Bube's brewery dug and expanded the natural caverns under the building to create large catacombs that were used for cold storage. Alois Bube died suddenly, just a few years before Prohibition, but his business had accumulated enough wealth that his family members simply closed the brewery, left everything intact, and made the hotel their home until the 1960s. Recently, the building was purchased and renovated to accompany a new craft brewery and restaurant. Visitors to **Bube's Brewery** can see the original broiler and smokestack in the Beirgarten, enjoy a meal in the Victorian hotel, visit the tavern in the original bottling plant, or descend 43 feet to enjoy a meal in the catacombs of the hand carved stone caverns.

Bube's Brewery

Date of Visit: _____

Sampled _____

_____

Comments: _____

_____

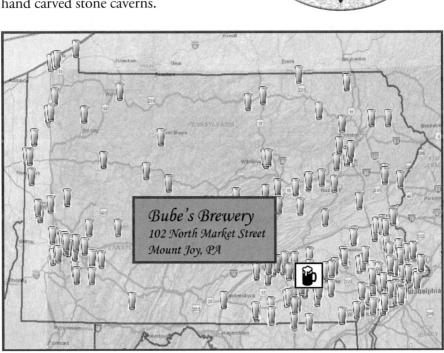

Bube's Brewery
102 North Market Street
Mount Joy, PA

One of the long-lasting, more famous brewers that emigrated from Germany to America eventually ended up in Pottsville. Initially established as Eagle Brewery, the company took the founder's Americanized name after a fire destroyed the original building and he joined forces with his son. **D.G Yuengling and Son** is listed as the oldest operating brewery in America, and they boast the longest uninterrupted history of management by a single family in the country. It is not uncommon to see the sixth generation Yuengling and his four daughters working around the factory and visitor's center.

D.G Yuengling and Son

Date of Visit: _____

Sampled _____

_____

Comments: _____

The location of the brewery was selected for the nearby spring that provided all the water for the beer and the adjacent steep mountain slope where tunnels were carved into the rock to provide natural cooling for the beer. Free, informative, and fun tours are given several times a day, and one of the first stops is a walk through the cool, damp caves. Visitors walk past the brick walls built by the government to block the caves during Prohibition before going upstairs to see the working tanks and bottling facilities.

D.G. Yuengling and Son survived Prohibition by producing porters for medicinal purposes, three different types of near-beer, and ice cream. When the 21st Amendment was ratified in 1933, the company created a Winner Beer and sent a truckload to President Roosevelt. One of the murals overlooking the larger brewing barrels displays the joy of opening a winner beer. In the years following Prohibition, the company launched an extensive modernization program to grow the business. The company continued to reinvest and update equipment and develop additional beers through the century that enabled them to become a nationally recognized brand.

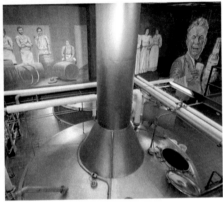

As immigrants filled the western lands of Pennsylvania, beer followed. This created a colorful history of the state and built a strong economic foundation. Frank Hahne immigrated to the United States from Germany. He learned brewing practices in a variety of cities, including Philadelphia and Milwaukee, before establishing his own company in the centrally located town of DuBois. The town was originally settled by the Germans and French as a lumber town, but the extraction of coal soon took over as the primary economic industry, filling the town with plenty of thirsty customers. The local reservoir and watershed provided ample clean water with which to brew beer. By 1906, the brewery had four products on the market, including a pilsner and porter, but most notably they produced DuBois Budweiser. When Anheuser-Busch found out about the name, they filed a lawsuit. A district court in Pittsburgh, and another appeal to a higher court, both favored The DuBois Brewing Company, who claimed ownership of the name "Budweiser." Frank Hahne Sr. passed away in 1932, leaving the company to his son, who eventually sold it in 1967, allowing the name "Budweiser" to finally go to Anheuser-Busch.[61]

In 2015, the **Doc G's Brewing Company** restored the tradition of brewing beer in DuBois when they renovated an old air conditioning and plumbing store and opened a family restaurant. Along with medical degrees, the owners, Drs. Jeff and Jen Gilbert, are avid craft beer connoisseurs and are committed to revitalizing Jeff's hometown. In honor of the original DuBois Brewing Company, they regularly serve G-wiser, a 5.5. ABV Czech style pilsner.

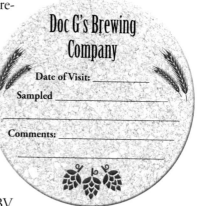

Just a little northeast of DuBois is **Straub Brewery Inc.**, located in the rural, mountainous community of St. Mary's between Allegheny National Forest and Elk State Forest. The city claims strong German heritage, and the brewery was founded in 1872 by one of the early German immigrants who passed the craft down to his sons. Similar to **D.G Yuengling and Son**, **Straub Brewery Inc** survived Prohibition by producing non-alcoholic and near-beer. The busi-

61. Wagner, Rich. 2012.

ness remained in the family for over 140 years and claim to use the same recipes and uphold the same environmental concerns that the business was founded upon. Today, it is recognized as an American Legacy Brewery that promotes a long tradition of Germanic style lager. Daily tours are provided free of charge Tuesdays through Fridays, and by special appointment on Saturdays; however guests must register or call several days ahead to ensure a limited spot. Tours end at the eternal tap where visitors can sample beer before entering the gift store.

As people moved into the western frontier, Fort Pitt was built and occupied by the British Army, who of course wanted and needed beer. The historical evidence of a brewery is limited, but several letters and personal journals report rations of beer to the army, and there are accounts of woman cooking and brewing beer. The first commercial brewery opened by the British in the area we now know as Pittsburgh was called The Point Brewery. When the fort was decommissioned, the salvageable materials were sold, and the buyer, Perter Shiras purchased Point Brewery brick to make a new brewery. Passed on to his sons, the brewery stayed in operation until 1835.[62]

One of the newer breweries in western Pennsylvania, **The Brew Gentlemen Beer Co.,** pays tribute to these early settlers in a town named for someone who undoubtedly would have regularly imbibed beer. Major General Edward Braddock was born in Scotland and rose to the ranks of Commander-in-Chief of the 13 colonies for Great Britain. Despite a tactical career, he is best remembered for the disastrous expedition in 1755, during the French and Indian War, where he and most of his men lost their lives.

The Brew Gentlemen Beer Co.

Date of Visit: _____

Sampled _____

_____

Comments:

_____

The namesake town of Braddock, where the disastrous battle occurred, is located along the Monongahela River several miles upstream from Pittsburgh. The town further developed its interesting history when a century later, Andrew Carnegie, established the Edgar Thomason Steel Works on the historic site of Braddock's Field. The town developed into a thriving industrial hub for the steel industry, which eventually crashed in the 1970s. Far from its heyday in the 1920s, the town suffered from unemployment, drug and crime problems in the 1980s and 1990s, and saw a 90 percent decline in its population. Since 2005, Braddock has made significant efforts to attract new residents and encourage creative and artistic communities. One of the major contributors to this revitalization is **Brew Gentlemen Beer Co.**

Matt Katase and Asa Foster founded the **Brew Gentlemen Beer Co.** They uniquely gain inspiration from Japanese philosophy and renovated a vacant electrical supply store with simplicity and elegancy. The large space is slightly separated by wide open doors that give the appearance of separated

62. Baron, Stanly. 1962.

spaces that blend together with the striking use of dark grey walls and sim-plistic artwork. The two owners learned to brew together during their collegiate years and have since refined their craft. Today, their flagship beers include General Braddock IPA and Carnegie Premium American Blonde Ale, along with a range of other beers including Mexican coffee oatmeal stout, Deep Breakfast coffee milk stout, Blanks and Postage English mild, and Mr. Automatic robust porter. **Brew Gentlemen Beer Co**. recently won an amazing accolade: in a blind taste test of 176 of the best double IPAs in the country, Paste Magazine ranked Brew Gentlemen's "Ales for ALS" number one![63] This amazing beer not only beat out top double IPAs from well-known breweries across the country for its taste, but it has a worthy cause with its proceeds supporting ALS research.

In 1861, Edward Frauenheim, a German immigrant, founded Iron City Brewery in the growing industrial center of Pittsburgh. He joined forces with other brewers and investors, most notably Anton Benitz and John Miller, and the brewery moved to a new location on Liberty Avenue and 34th Street. The merger led to the development of several large buildings with an unprecedented brewing capacity that made them the largest brewery in Pittsburgh. Mergers continued with other regional breweries, and they took on the new name: the Pittsburgh Brewing Company. By 1899, the Pittsburgh Brewing Company was the largest brewing operation in Pennsylvania and third largest in the country.[64] They survived Prohibition

---

63. Vorel, Tim. 2017.
64. Wagner, Rich. 2012.

by producing near-beer and ice cream and ran a cold storage business. Post-Prohibition, their success continued for the next century with the enormously popular Iron City beer, and later, I.C. Light. Eventually the company declined, filed for bankruptcy, and moved brewing to Latrobe. The building is registered as a Pittsburgh Historic Landmark.

Around the same time as the establishment of Iron City Brewery, the Duquesne Brewing Company established a brewery on the corner of Mary and 21st Street, in Pittsburgh. One of its major innovations was the use of refrigerated box-train cars that transported beer to western states, which was a novel idea in the early 20th century. In 1900, it was the first brewery in the nation to use an electric truck and was the first brewery to pasteurize their beer bottles. Other technical advances included the use of mechanical refrigeration and steam heat. With the intent of competing economically with The Pittsburg Brewing Company, it consolidated multiple brewing facilities to fall under a single ownership. Fifteen small breweries surrounding Pittsburgh joined forces with the Duquesne Brewing Company to form the Independent Brewing Company (IBC) of Pittsburgh in 1905. The Duquesne Brewing Company remained one of the leading branches within the IBC, and following the end of Prohibition in 1934, a name change back to the Duquesne Brewing Company reflects its importance. Along with the **D.G Yuengling and Son**, they too expressed their appreciation by sending President Roosevelt a case of beer as a gift for signing the 21st Amendment.

The size of production and strength of heritage allowed many Pennsylvania breweries to survive Prohibition and economically compete with the growing industrialized breweries in the midwest for several decades. In 1946, a survey conducted in Philadelphia, the country's third most populous city at the time, ranked preferences of the United States' top leading beers. Of the 19 beers, nine of them were produced in the city of Philadelphia and another five were produced elsewhere in the state of Pennsylvania. Schmidt's and Ortlieb's were at the top of the list and preferred nearly twice as much as Anheuser-Busch.[65]

Despite Philadelphia's large population having a preference for local beer, competition was tough. Eventually, most breweries in the city closed their doors or sold their facilities to the industry's giants, with the Hornung Brewery and the Hohenadel Brewery folding in 1953. The Gretz Brewing Company tried to avoid foreclosure by active marketing. They sponsored both men's and women's bowling teams and a televised beauty pageant, but in 1960 they were forced to sell their brands to Esslinger's Inc., who lasted a few more years but eventually closed in 1964. The Ortlieb Brewery had been passed down as a family business for generations. The enormous building purchased in 1899 encompassed an entire city block on 3rd street and had multiple brands and financial interests in several other Pennsylvania breweries. They too sponsored many events in Philadelphia and were well known for charitable donations to the March of Dimes and the American Cancer Society. They finally sold their brands to the C. Schmidt and Sons Brewing Company in 1981.

---

65. Wagner, Rich. 2012.

Schmidt beer was a popular product in Philadelphia for over 125 years. This longevity became part of their promotion slogan, "None Better Since 1860." The company expanded by buying several breweries and bottling facilities in Philadelphia, Norristown, and Cleveland and they entered the modern era by leading in technological advances in mass production. They became Pennsylvania's largest brewery and distributed throughout the United States, but finally succumbed to the pressures of a failing business and closed their doors in 1987.

For the first time in over 300 years, Philadelphia was a city with no active breweries. Ironically, the 102nd meeting of the Master Brewers Association of the Americas was hosted in a city with no breweries, but a glimmer of hope was on the horizon. In 1989 and 1990 a few brewpubs appeared across the state. Although a few companies such as Red Bell, Independence, and Poor Henry's had limited brewing success and were forced to close within a few years of opening, they set the momentum for an industry change. Secretly waiting, experimenting, and refining their craft, an underground movement of homebrewers were ready to explode and show the world their LOVE OF BEER.

CHAPTER 4

The Craft Beer Explosion

The American brewing industry has changed drastically in the past few decades. The large multinational conglomerates that fall under the more commonly known names of Anheuser-Busch, Miller, and Coors dominated the industry in the second half of the 20th century. In order to compete with each other, they reduced costs by adding corn and rice, making their beers lighter in calories, and almost indistinguishable in flavor. Blind taste tests done in 1964 and 1971 found consumers could not distinguish between brands.[66] Thanks to the legalization of home brewing by President Jimmy Carter in 1979 with H.R. 1337, followed by the legalizations of brewpubs by individual states in the 1980s and 1990s, craft breweries exploded.[67] Only eight craft breweries existed in the United States in 1980 to well over 6,000 today, and the industry is still growing. Americans can boast that 98 percent of all our breweries are considered small and independent.[68]

---

66. Caroll, Glenn R. and Anand Swaminatham. 2000.
67. McLaughlin, Ralph B, Neil Reid, and Michael S. Moore. 2014.
68. Brewers Association. 2017.

One great example is the **Lancaster Brewing Company,** which spearheaded the resurgence of breweries in the Lancaster area and is trying to relive the glorious beer drinking past of the early 1800s when Lancaster County produced nearly 7 percent of all the beer brewed in the United States. The county was established in 1730 when residents of Chester County complained that "thieves, vagabonds, and ill people" occupied the western portion of the county and forced them to secede. English, German, and Scot immigrants settled the new county, and it became a thriving community. The city of Lancaster served as the nation's capital for one day when the Centennial Congress fled Philadelphia in 1777, and it served as Pennsylvania's state capital from 1799 to 1812. Tobacco was the leading cash crop for most of the 1800s

Lancaster Brewing Company

Date of Visit: _____

Sampled _____

_____

Comments: _____

_____

and the silk factories employed thousands of workers. It is not surprising that numerous taverns and breweries existed.

Lancaster Brewing Company serves excellent beers and unique food in the historic Edward McGovern Tobacco building where customers sit in a two-story open building with a rustic ambiance. The open floor plan with exposed wood beams and brick walls revolves around an antique freight service elevator and allows people to look down on the brewing tanks.

Although mass-produced beer still holds the market for volume of sales, a craft brewery movement is changing the industry. Budweiser sales peaked in 1988, producing 50 million barrels of the popular beer a year, accounting for a quarter of the U.S. beer market.[69] People still love Budweiser. It's an iconic American symbol that remains one of the most popular beers in the United States, but its sales have declined by 68 percent.[70] This decrease is not because people stopped drinking beer; they just started to drink different kinds of beer. Over the past decade, while mass produced beers were declining, craft beer sales were increasing. In 2016, craft breweries sales increased 6.2 percent, which accounted for 12.3 percent of dollar sales with a predicted increase of 7 percent annually for the next five years.[71]

69. Notte, Jason. 2016
70. Felberbaum, Michael. 2015
71. Brewers Association. 2017.

Weyerbacher
Brewing Company

Date of Visit: _____

Sampled _____

_____

Comments: _____

_____

The **Weyerbacher Brewing Company,** located in Easton, is a great brewery that provides an enlightening tour where visitors learn about the history of craft brewing, fun facts on how their beer is made, and how-to suggestions for sampling craft beer. Originally founded in 1995 in a nearby livery stable, the company moved into a larger production facility plant with a large bar, open seating, and an outdoor patio. Several free, daily tours are provided. They last about 30 minutes, are informal, and guests are encourage to bring a drink with them.

Weyerbacher Brewing Company established its reputation early as a craft brewery that went against the generic, mass-produced beer, and dedicated themselves to creating exciting, full-flavored beers. Termed "Big Beers," Weyerbacher excels at creating high-quality brews that typically have a much higher alcohol content, which helps to explain some of their well-known beer names, such as Blithering Idiot (11.1 percent ABV) and Merry Monks (9.3 percent ABV). A few of the beers with names that connect the brewery to their local consumers include Last Chance IPA, where proceeds go to local animal rescue organizations, and a September release for NFC football fans called Dallas Sucks.

Across the United States and Pennsylvania, craft breweries contribute to local, regional, and national economies. In 2016, the industry is estimated to create nearly half a million jobs and $68 billion in wages and benefits a year across the United States.[72] The benefits extend past those jobs directly in the breweries into related industries such as construction, retail, and government. Additionally, locally-owned breweries tend to make positive impacts on their communities. Money spent at locally-owned businesses tends to stay in the community more so than when spent in chain stores, and the increase of local, small businesses often correlates to an areas' economic growth.

Across the United States, the location and distribution of craft breweries is somewhat uneven. Although there is at least one in every state, and most Americans live within 10 miles of a craft brewery, the largest concentration is in California, The Pacific Northwest, and Colorado.[73] A high concentration of German descendants in Wisconsin, Minnesota, and Michigan creates a secondary hub of craft breweries near the Great Lakes. A third concentration of breweries is found in the historic beer-producing region of the mid-Atlantic, particularly Pennsylvania. Religion, legalization, and lack of home brewing are often cited as reasons for the slower growth of breweries in the south. Across the country, typically the largest concentration of breweries are primarily in areas with high population; however when normalized for total population, states such as Vermont and Florida

72. Brewers Association. 2017.
73. Hoalst-Pullen, Nancy, et al. 2014.

and specific cities such as Ashville, North Carolina, and Myrtle Beach, South Carolina tend to have higher concentrations of breweries relative to their population, probably due to tourism.[74]

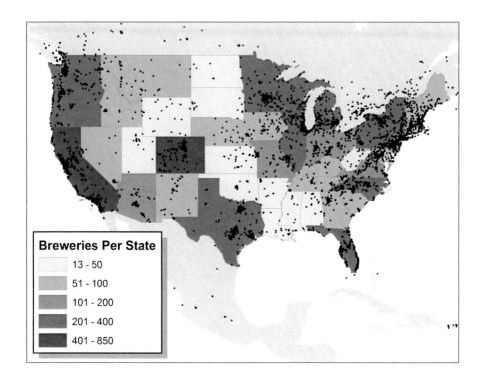

The increase in craft breweries coincides with a change in Americans who are reacting against mass-produced, generic products and are looking for self-expression, particular through higher quality goods.[75] The sentiments are part of the larger neolocalism movement, defined as the conscious attempt of individuals and groups to re-establish local ties. It is the deliberate seeking of local attachment by residents who demand local goods and services. It often implies uniqueness with ecologically sound, sustainable practices that tend to empower self-sufficient people, support local economies, and build social networks. Similar changes are seen with an increase in wineries, farmer's markets, community gardens, and farm-to-table initia-

74. Baginski, James and Thomas L. Bell. 2011
75. Schnell, Steven M. 2011

tives. Terrorism, war, and global economic meltdowns are often cited as reasons for an increase in people's desires for an idealized nostalgic past. Additionally, with an ever-increasing highly urbanized population that has to deal with noise, city stress, traffic, and crime, people are nostalgic for a rural idyllic setting and develop an affection for a vanishing quiet, rural countryside and simpler way of life.

Undoubtedly, craft breweries have catered to the movement by marketing a romanticized past, emphasizing unique identities, and creating a truly local experience. Breweries are proud to promote their idiosyncrasies, marketing their local place as much as they are their beer. Many breweries are contributing to the cultural heritage by reusing old buildings that are commonly decorated with historic maps and photos. Exposed beams, old tools, and brick walls reinforce authentic heritage. Local pride and the promotion of a true craft, workmanship, and tradition are visible throughout the consumption process. Names of beers typically come from historic events, local legends, and wildlife, and brewers go to great lengths to create dis-

tinctly local themes while images adorn the beer labels and drinking establishments. These unique settings clearly match the brewery's identity to
their storefront location. For example, the **Breaker Brewing Company** in
Wilkes Barre is located in an old brick building
that once housed a school. It sits on the
geologically significant steep slopes of a
double plunging syncline that overlooks the Wyoming Valley in an economically rich anthracitic coal field.
The decor and menu items honor the
heritage of the coal workers and
beers called Lunch Pail Ale, Rye-Om-
ing Valley IPA, Olde King Coal Stout,
and Minefire Blackberry Jalapeno Ale
whimsically reference the legacy and culture of the area.

Breaker
Brewing Company

Date of Visit: _____

Sampled _____

_____

Comments: _____

_____

Most breweries publish statements on their menus, webpages, and social media stating their connections with local goods and services. Many breweries buy food from local farmers, display art by local artists, host local musicians, and donate to local charities. Many are pairing up with wineries or other breweries to develop a gastronomic region for tourism. Although most craft breweries formed after 1980, they create the facade of an old industry. Many reside in unique, historic buildings, and contribute to gentrifying neighborhoods within cities. Most craft breweries are conscious of their role in society and implement various sustainability measures.

Craft beers are important in consumer culture for they tend to attract a certain demographic. The different flavors and styles of beer, along with the unique establishments to drink them in are developing a complex, cultural phenomenon with a sophisticated, prestige factor. The average consumer of craft beer is a well-educated, white male earning an average income of over $75,000 a year.[76] Typically between the ages of 21 and 44, this group, millennials, are born after 1980, and are the prime market for the beer industry. Millennials are 34 percent more likely to drink beer than the average American, and they are 26 percent more likely to buy craft than mass-produced beer.[77] Additionally, they lean towards strong environmental sentiments, are more adventurous, and are willing to seek out, travel farther, and pay slightly more for a more unique, high-quality experience. And as the prime target audience for craft brewers, millennials follow social media and are five times more likely to be influenced by word-of-mouth than advertising.[78]

76. McLaughlin, Ralph B, Neil Reid, and Michael S. Moore. 2014.
77. Notte, Jason. 2016.
78. McLaughlin, Ralph B, Neil Reid, and Michael S. Moore. 2014.

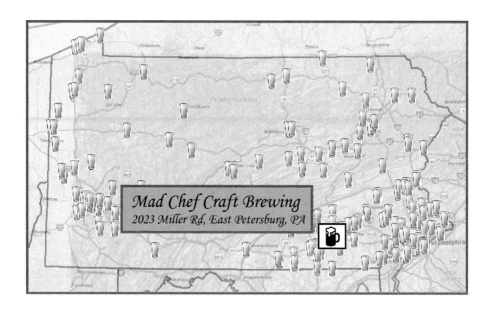

An example of social media's influence is seen on the chalkboard of **Mad Chef Craft Brewing Inc**. Located in East Petersburgh, this brewery has a more modern, industrial-feel for serving beers. Simple but elegant food sets this brewery apart. The two owners have many years of experience in the culinary industry, including a degree from the Culinary Institute of America. Served in a friendly, relaxed atmosphere alongside beers with whimsical names, such as "Cherry on My Wayward Son."

Mad Chef Craft
Brewing Inc.

Date of Visit: _____

Sampled _____
_____

Comments: _____
_____

The big beer companies have begun to recognize the impacts of these craft breweries and their ability to attract a small but desirable demographic of young, educated, and affluent beer drinkers. In attempts to compete, faux microbrews have been developed with minimal connections to their larger parent companies.[79] CoorsMolson developed Blue Moon, Miller introduced Leinenkugel, and Anheuser-Busch produced Shock Top. Anheuser-Busch introduced two organic beers under the labels Crooked Creek Brewing Co. and Green Valley Brewing Co. In addition, many of the large companies started to buy smaller craft breweries. Anheuser-Busch bought one of the oldest craft breweries in the United States, Blue Point Brewing Company, in February, 2014, along with the acquisition of other prominent craft breweries such as Goose Island, Elysian, Blue Point, 10 Barrel, Golden Road, Four Peaks, and Breckenridge over the past five years.

Not surprisingly, the marketing campaigns by the big beer companies have targeted this consumer culture. Miller Lite reintroduced their retro style beer can with a look that implies tradition, and sales immediately increased. They developed a series of ad campaigns "Miller Lite. We invented light beer and you[80]," along with another advertisement with a farmer examining cascading hops similar to craft beers. Nowhere is this competition more apparent than in the 2015 Super Bowl ad. After several decades of losing ground to craft brewers, Anheuser-Busch recognized it may not be possible to win back those customers and instead marketed to their core consumers.[81] The recent campaign says Budweiser '...isn't brewed to be fussed over.' The 2015 Super Bowl ad ran as a series of images with bold words stating that Budweiser is proudly a macro brew and that it is brewed the hard way. They reinforced tradition by showing images of old advertisements selling Budweiser for 15 cents and stated that it has been brewed the same way since 1876. By contrast, they showed serious men sniffing small glasses of craft beer followed by images of a fun partying crowd drinking Budweiser. A complete reversal occurred in the 2016 Super Bowl, where a 30-second

79. Schnell, Steven M and Joseph F. Reese. 2014
80. Kell, John. 2015.
81. Felberbaum, Michael. 2015

ad cost $5 million, up from $2.5 million a decade earlier.[82] For the first time during the Super Bowl, Anheuser-Busch InBev promoted Shock Top, a product they adamantly defend as a craft beer.

But let's face it, local beer does taste better; in fact, it tastes great. If drinking a beer was all about image and prestige, the large companies would have seamlessly unlimited funds to sway consumers. Sit in any craft brewery, and you will find it does not take long before you overhear a local person saying they were anxiously waiting for a new seasonal beer to be tapped or travelers talking about their journey — sometimes deliberately seeking a new brewery — and ordering a sampler flight. Often the excitement will be generated by the head brewer or staff as they eagerly add to the chalk board that a standard recipe has been tinkered with and the new nitro variation is smoother and creamier. Usually within sight of the bar is the heart of the operation: kettles and fermenting tanks shimmering after

---

82. Notte, Jason. 2016.

the day's brewing and the smell of hops wafting through the air. It is a social and cultural experience, and it is the variations that generate a unique atmosphere that cannot be manufactured in ads by the large corporate giants.

Luckily in Pennsylvania, we do not have to travel too far to find a great local brewery. The state claims over 350 breweries with no signs of slowing down. From only a few breweries in the mid-1990s, a slow but steady growth occurred in the early 2000s, followed by a rapid expansion in the 2010s. In fact, the number of breweries doubled in the past four years. The overall distribution of breweries in Pennsylvania follows the national trends.[83]

---

83. Feeney, Alison E. 2015

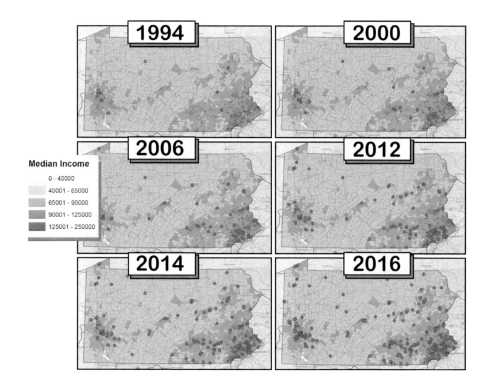

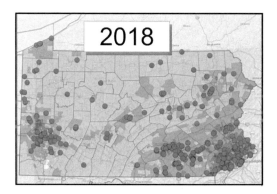

The majority of early breweries were located most often near large population centers. Originally, many started gentrifying declining neighborhoods in large urban cities, most notably Philadelphia, Pittsburgh, and Harrisburg. For example, The **Appalachian Brewing Company** opened in Harrisburg in 1997 and was the first brewery in the state's capital in 46 years.

Appalachian Brewing Company

Date of Visit: _____

Sampled _____

_____

Comments: _____

_____

The site was chosen in part for the incentives offered by the city's downtown development authority. The building's history dates to the 1890s when it was owned by the Harrisburg Passenger Rail Co. and Harrisburg Trolley Works, but was greatly enlarged in 1918 to house a machine print shop. Several fires and several decades of storage use took its toll on the building, but it was finally acquired in 1995 and heavily renovated for two years to make it into the large, beautiful, thriving brewpub it is today. Renovations utilized as much of the brick walls, heavy timber beams, and hardwood floors as possible, which despite its large size, still provides a warm cozy feeling.

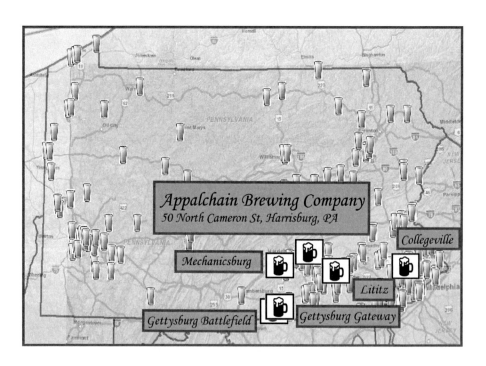

Appalchain Brewing Company
50 North Cameron St, Harrisburg, PA

Collegeville

Mechanicsburg

Lititz

Gettysburg Battlefield

Gettysburg Gateway

Appalachian Brewing Company has expanded and now brews in six locations. The Collegeville, Lititz, and two Gettysburg breweries have smaller 4.5-5 barrel brewing systems that often brew seasonal and specialty batches, along with flagship beers; in Mechanicsburg, the larger 15-barrel system brews beers along with their line of popular soda. Additionally, "The Abby Bar" in the upstairs of the Harrisburg location has evolved into a well-established music and entertainment venue. Their newest brewpub is now open in Shippensburg.

Manayunk
Brewing Company

Date of Visit: _____

Sampled _____

_____

Comments: _____

_____

The next wave of breweries in Pennsylvania tended to locate in the more trendy suburbs, establishing themselves in unique and distinct buildings. Manayunk was one of Philadelphia's first neighborhoods to undergo revitalization. Located just a few miles west of downtown, it historically housed an important textile industry along the Schuylkill River, but as factories closed the neighborhood declined. Today, numerous restaurants, bars, and rail trails bring people in by train station or Interstate 76. The **Manayunk Brewing Company** is celebrating 20 years of brewing. Located in an 1800s textile mill, antiques and scales used in the wool factory adorn the restaurant where customers can imbibe Schuylkill Funk Ale, Belgian Style Yunker's Nocturnum, and Brotherly Love Hard Apple Cider, all with the company's logo depicting the historic, iconic limestone bridge found at the mill.

As the number of breweries increase and the industry matures, many of the more recently established companies are located along main streets in small towns and contribute significantly to local communities throughout the state. Breweries are bridging ties with other local business, contributing to local development, and promoting local heritage.

Most Pennsylvanians live close to a local brewery or two, sometimes even 10. The largest number of breweries per county is located near big cities, but when normalized for total population, several counties in the north, both in the eastern and western portions of the state, have higher proportions of breweries. These counties are known for their outdoor activities, tourism, and natural resources. In the south-central part of the state, Adams County is known for the popular destination Gettysburg and has a surprisingly high proportion of breweries compared to its population. Similarly, craft breweries typically attract a specific demographic market that correlates to a well-educated, higher income consumer. Many of the college towns, such as State College, Edinboro, Bloomsburg, and Mansfield, appear to accommodate more craft breweries than their smaller population would likely support. Statistically, total population, median income, and percent of residents holding a college degree accounts for the most significant influences on the presence of breweries across the state.

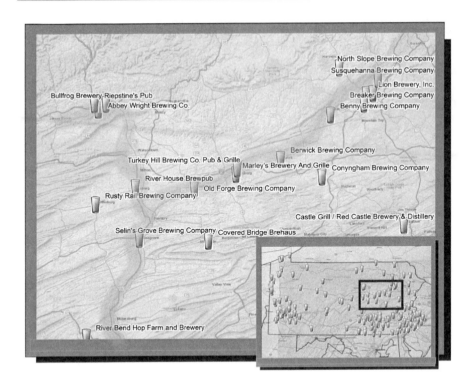

Just as in the past, transportation corridors and mining seem to have a close connection with breweries, and Pennsylvania is well known for its hub of highway networks and its vast extractive resources. River towns were historically an integral part of Pennsylvania's commercial and industrial economy. The location of abundant natural resources combined with plentiful waterways led to the development of many bustling towns, followed by rail lines and later highways. With the decline of some extractive resources, many river town communities today have struggling economies. Several breweries are revitalizing these towns, promoting, preserving, and enhancing their heritage.

Berwick
Brewing Company

Date of Visit: _____

Sampled _____

_____

Comments: _____

_____

Berwick Brewing Company is an excellent example of a craft brewery that has helped revitalize an industrial town. Located right along the Susquehanna River, they renovated an old bakery and provide a family atmosphere for the community to gather. Large open rooms, a bier garden, and great pizza keep the hometown feel in this unpretentious brewery.

Several craft breweries are located in traditional mining towns throughout the state. Carbondale, for example, is located in Lackawanna County in the heart of coal country. It is often referred to as "The Pioneer City" because it was the site of the first deep anthracite coal vein that led to the regions' lucrative mining industry. The town attracted immigrants from Wales, England, Scotland, and Ireland, and architecturally impressive buildings, such as the city hall and courthouse and the large post office, recorded the town's success. **Iron Hart Brewing Co.,** owned and operated by Matt Zuk, evolved from the former brewery called 3 Guys and a Beer'd and helps to keep local businesses downtown. It sits on the corner of Church Street and Lincoln Ave and offers a cozy neighborhood pub with eight beers on tap.

Iron Hart
Brewing Co.

Date of Visit: _____

Sampled _____

_____

Comments: _____

_____

Susquehanna Brewing Company

Date of Visit: _____

Sampled _____

_____

Comments: _____

_____

Approximately 30 miles southwest of Carbondale is the town of Pittston. Coal was first discovered in Pittston in the 1770s, but the town grew after a canal was built in the 1830s and later the Lehigh Valley Railroad in the 1850s. The owners of the **Susquehanna Brewing Company** replicate recipes that their grandfathers' grandfathers brewed in the region years ago. In an interesting timeline, the owners describe the history of Charles Stegmaier who emigrated from Germany and started brewing in Northeast Pennsylvania. Over the years, the company changed hands and entered partnerships with other brewers, but eventually in 2010, the **Susquehanna Brewing Company** was formed and brought artesian beers back to the region.

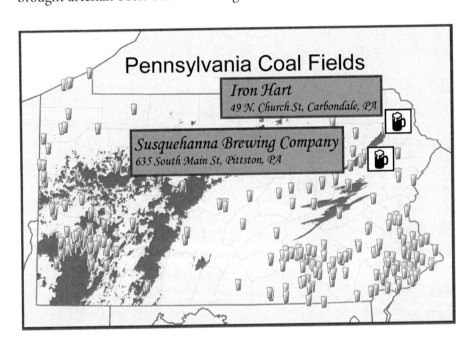

Pennsylvania Coal Fields

Iron Hart
49 N. Church St, Carbondale, PA

Susquehanna Brewing Company
635 South Main St, Pittston, PA

Throughout Pennsylvania, traditional mining towns grew in the late 1800s, mostly from the large number of German immigrants who came to work in the manufacturing industries. Originally settled as a sawmill town for the lumber industry, Titusville is commonly recognized as the birthplace of the modern oil industry, which expanded to include related steel, iron, and rail industries. **The Blue Canoe Brewery** was established in 2008 and is a central fixture in the heart of downtown Titusville. A mural on the outside of the building depicts the natural resources and history of the town, and inside a large open room provides customers view of the canoe mounted over a large bar listing the beers on tap.

The Blue Canoe Brewery

Date of Visit: _____

Sampled _____

_____

Comments: _____

_____

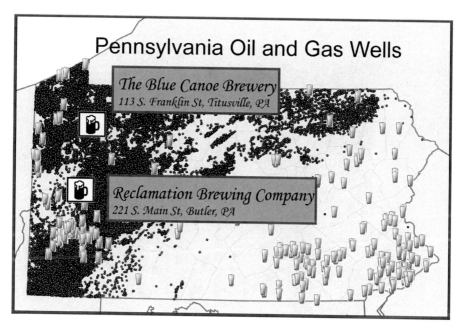

Pennsylvania Oil and Gas Wells

The Blue Canoe Brewery
113 S. Franklin St, Titusville, PA

Reclamation Brewing Company
221 S. Main St, Butler, PA

Many breweries are located near Pennsylvania's active oil and gas fields, and one called **Reclamation Brewing Company** certainly reflects that. The three owners are dedicated to preserving the local community, providing a family dining experience, and educating their customers on beer by reclaiming the beer drinking experience from the negative Prohibition-era views and the bland tasteless encounters of the past.

Reclamation
Brewing Company

Date of Visit: _____

Sampled _____

_____

Comments: _____

_____

Craft breweries across Pennsylvania embrace their local community and proudly promote through their websites and social media, providing details about the building location, type of establishment, owners, brewers employed, names of beers, what's on tap, and community events. Words such as passion, craft, fresh, and neighborly are commonly used to express the commitment to the local community in providing a quality product. Some breweries go as far as to state ethical behavior, honesty, and integrity as fundamentally important to the business as the beer that they brew.

Boneshire Brew Works is one of the state's new businesses and is an excellent example of how breweries are making a unique and significant contribution to local communities. Alan Miller, renovated a vacant old machine shop and within the first year of business built an active, thriving business. On any given night, an eclectic group of customers can be found sharing a beer. Located half way between Harrisburg and Hershey the clientele is as diverse as the brewery's activities which range from live bands to movie night, to trivia night, to game night with Jenga and Connect Four. All the activities are updated on their Facebook page, which has thousands of followers.

Boneshire
Brew Works

Date of Visit: _____

Sampled _____

_____

Comments: _____

_____

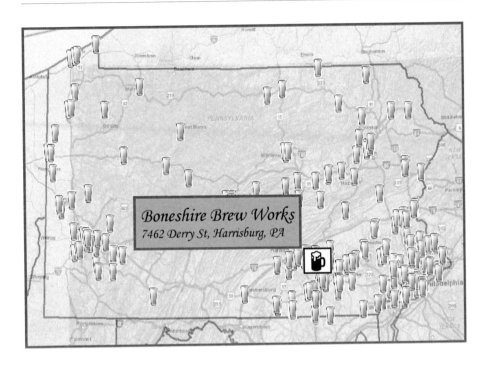

Boneshire Brew Works exemplifies how craft breweries help restore neighborhoods in large cities and redefine smaller communities. Along with re-using the buildings and adapting the space to fit a brewery, **Boneshire Brew Works** acquired, recycled, and refinished the wood from the shipping crates of a new, larger brewing system purchased by a nearby brewery, **Al's of Hampden/Pizza Boy and Co.** The recycled material was used to build the long bar and focal point of the brewery. Additionally, like many new craft breweries, **Boneshire Brew Works** is making strides to package beer in economically and environmentally conscious cans and support local farmers by purchasing hops and other essential ingredients. And despite the extremely long, hard hours Alan and his staff put in, he has a wonderful smile on his face as he pours his latest recipe to make sure everyone embraces the LOVE of BEER.

CHAPTER 5

Craft Breweries and their Great Buildings

One of the most enjoyable aspects to drinking a craft beer is consuming it at the brewery. The beer is extremely fresh, the brewers are usually hanging around the bar to answer any questions or ask customers their opinions on flavors, and the buildings and their environs are usually a continuation of the craft. The buildings, along with the bar, taps, tables, stools, and decorations, are often an extension of the personalities of the individuals making the beer and their connections with the local community. Although most of the breweries in Pennsylvania were only established in the last few years, they tend to reuse pre-existing buildings, portray historic images on their labels, and overtly promote the heritage of their drinking establishment.

Architecture is an important part of the landscape that tells the story of past cultures and peoples. It tells how the environment shaped how they lived, worked, and played. America lacks the great architectural heritage of European countries, so it is even more important to preserve historic structures. People appreciate older buildings, and many individuals and organizations are working hard to preserve representative samples from the past that will instill a deeper appreciation for people's surroundings.[84] Programs such as The National Historic Preservation Act and The National Register of Historic Places were established to protect unique buildings that have historic characteristics. Additionally, individuals and local communities have developed their own programs to help us appreciate our past.

In an era of global businesses and massive chain restaurants that are often managed and operated by a few large parent companies, patrons can expect to find the same menu, décor, and service regardless of where they travel.

---

84. Merlino, Kathryn Rogers. 2014.

Although familiarity and comfort coincides with that expectation, cookie-cutter strip malls and chain stores are making many suburban areas and small towns appear similar. History can be a powerful tool in community relations and should be utilized in local development. Heritage has proven itself to be a lure for many patrons desiring authentic experiences. Local vernacular buildings that embrace the layers of use over time have an enormous potential for economic development, and in turn, create attractive places to live. Breweries with their ever-changing beer menu and food options, often sourced from local farmers, are the ideal type of small business that can embrace a city's uniqueness.

In Pennsylvania there are endless examples of breweries that preserve important architectural buildings and occupy distinctly unique spaces. With the adaptive reuse of these older buildings that are made accessible to the public, breweries restore cultural heritage, celebrate the legacy of local history, and pass stories onto future generations.

Several breweries occupy buildings listed on the National Register of Historic Places. One of the early pioneers of Pennsylvania's craft brewing movement set an outstanding example by renovating a Nationally Registered building in Pittsburgh. **Penn Brewery** was founded in 1986, and with assistance from the Pittsburgh History and Landmarks Foundation, they restored and occupied the 1848 Eberhardt and Ober Brewery. Located in the Troy Hill neighborhood that was originally settled by German immigrants who worked in the breweries and tanneries, the brewery overlooks the Allegheny River.

Penn Brewery

Date of Visit: _____

Sampled _____

_____

Comments: _____

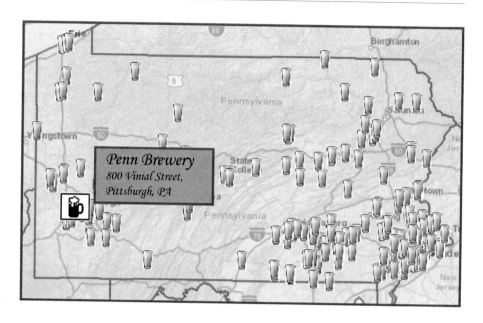

Five of the remaining buildings located on corner of Vinial Street and Troy Hill Road were designed in the 1880s by architect Joseph Stillburg. The brick castle-looking buildings are massive. The brew house has three-foot thick concrete and steel floors and a five-foot thick interior wall. Stone tunnels and caves, originally used for chilling lager, can still be found underneath the buildings. Connecting with the past, **Penn's Brewery** specializes in German-style beers that are brewed in the traditions of the immigrants who originally built the building and developed the surrounding Deutschtown neighborhood. Today, they serve their customers in the restored original keg-filling area of the old brewery.

Another example is The Tap Room (**Spring House Brewing Company**) that resides in the Nationally Registered Hager Building in downtown Lancaster. This five-story, French Renaissance Rival style building was designed and built by architect C. Emlen Urban in 1910, and it served as Lancast-

Spring House Brewing Company

Date of Visit: _____

Sampled _____

_____

Comments: _____

_____

er's prominent downtown department store. The Tap Room is located on the ground floor of this beautiful, ornate building amongst other commercial businesses. Large open windows allow customers to appreciate several neighboring architectural gems while enjoying their beer. In one direction, customers can view the Central Market Place or look across the street and gaze on the Nationally Registered Steinman Hardware Co building. Built in 1886, this brick and cast iron ornate building with stained glass windows claims to be the oldest hardware store in the United States.

The **Spring House Brewing Company** has opened a new facility just a few blocks down the street from the Tap Room's downtown location where they now brew and can all of their beer. Investing over $2 million in recent renovations of a 1901 warehouse, they have created a delightfully unique dining experience where layers of the past are proudly displayed.

Another building that houses a brewery listed on the Unites States National Register of Historic Places is located in Selinsgrove, Pennsylvania. Selinsgrove is considered to be one of the first white settlements west of the Susquehanna. Founded in 1713 by George Gabriel, the town was probably chosen for its ideal geography between Penn's Creek and the Susquehanna River. Nearly 100 years after settlement, the town's most illustrious citizen moved to Selinsgrove: Simon Snyder. Snyder was a self-educated man who viewed the move to the frontier as an opportunity. He bought land and co-owned a mill and general store with Anthony Selin, for whom the town is named. Snyder quickly gained local support in politics and became Pennsylvania's only three-term elected governor; he served from 1808-1817. Snyder was well known for his advocacy of free public schools and believed democracy could not thrive without educated citizens. Snyder was also the first governor of any state to officially speak out against slavery.[85] In his message to the state legislature on December 5, 1811, Snyder argued against the institution of slavery and stated that "all men are born free and equal."

---

85. Selinsgrove. 2017.

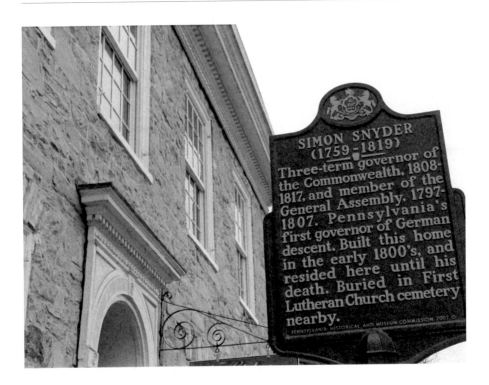

Simon Snyder built a mansion in 1816 on Market Street, which is now listed on the National Register of Historic Places. The **Selin's Grove Brewing Company** opened in 1996 and occupies the lower-level of Snyder's beautiful limestone mansion. The brewery is located in the original kitchen. It is small but quaint, with exposed rock walls and two large fireplaces. The brewery's logo of a dog inside a wheel reflects local history: an old brewery and distillery in Selinsgrove used the labor of stray dogs to generate power. In memory of working canines, the brewery donates yearly to animal shelters.

Selin's Grove Brewing Company

Date of Visit: _____

Sampled _____

Comments: _____

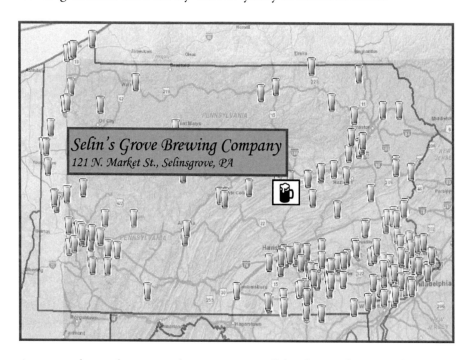

Selin's Grove Brewing Company
121 N. Market St., Selinsgrove, PA

Anyone who truly enjoys the experience of drinking a beer needs to visit the **Battlefield Brew Works** in Gettysburg. Amongst Gettysburg's vivid history that visitors encounter throughout the National Park and the surrounding town, **Battlefield Brew Works** has authentically and effectively established a family-friendly environment in a large barn that is listed on

Battlefield Brew Works

Date of Visit: _____

Sampled _____

_____

Comments: _____

_____

the Historic Barn Registry of Adams County. The adaptive reuse of this 1849 barn lets visitors experience the important Monfort Farm barn, which served as the hospital for the Confederate Army during the Civil War.

During the Civil War, field hospitals were established near battlefields to treat the sick and wounded. Private residential homes and barns were commonly used, and any available bed, chair, or table provided a raised surface to treat victims. Although the battle of Gettysburg only lasted a few days, the sick and wounded remained in the area for several months. The Monfort Farm was one of the Confederate Army's largest hospitals during the Civil War, serving over 1300 soldiers.[86] 47 soldiers died and were originally buried on the farm, but were later moved to Richmond, Virginia. Conditions were unsanitary, with gruesome surgeries performed near windows for available light, and the amputated limbs were tossed into a pile.

Patrons to this brewery not only have the opportunity to reflect on the significance of the battle, but they also experience a culturally significant feature found across the rural Pennsylvania landscape: the Pennsylvania

---

86. Civil War Photo Album. 2017.

barn. The barn is one of the state's important vernacular buildings whose architectural style was dispersed with early settlers throughout North American from 1790 to 1900.[87] The importance of agriculture to early settlers who built enormous barns while living in small log cabins can be seen in this type of banked barn. The barns were working, practical, and functional structures that heavily focused on grains and livestock. Built into a hill slope, the banked barn ensured easy access to both the basement and ground floor, with the most distinguishing feature being a forebay, or an overshoot of walls beyond its foundation.[88]

Entrance to **Battlefield Brew Works** is through the original ramp of the banked barn, and the pub inside sits within the large, open room with enormous wooden-supported rafters and wide-planked flooring. Decorative openings in the brickwork allow light to penetrate the room and the colonial plaster mix composed of mud and horsehair authentically crumbles off the wall. Simple and stately Civil War memorabilia decorate the environs, and visitors can look down on the large copper brewing system. Downstairs the entire brewing operation is located where the livestock would have been kept, and the original lime coated beams are still visible. In addition to beer, **Battlefield Brew Works** serves unique pub fair and has expanded to include a high quality distillery. The company vehicle is a hearse with "spirits on board" written on the side promotes the distillery with a fun play on words,

---

87. Drake, Dawn. 2009.
88. Ensminger, Robert. 1992

Two Rivers Brewing Company

Date of Visit: _____

Sampled _____

_____

Comments: _____

_____

Although not officially listed as Historic Places, many breweries occupy unique buildings dating from the 1800s and early 1900s that provide character and maintain the communities' past cultural heritage. For example, The **Two Rivers Brewing Company** is located in Easton, Pennsylvania and has renovated the Old Mt. Vernon Hotel built in the 1800s. Easton is located at the confluence of the Delaware and Lehigh Rivers and is equidistant from Philadelphia and New York City. It became an important commercial and transportation hub with easy access to Bethlehem's steel production and Northeast Pennsylvania's anthracite coal deposits. The Mt. Vernon Hotel was built to accommodate the bustling city's travelers, and it operated continuously as a hotel from 1855 to 1994, although illicit activities and a brothel probably supported its revenues over the years.

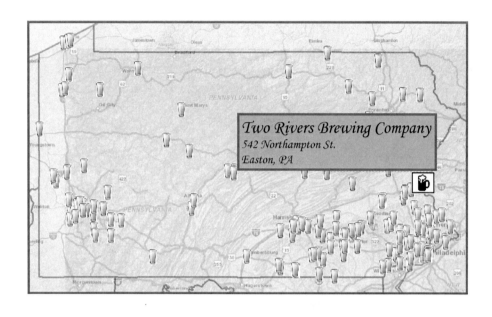

Two Rivers Brewing Company
542 Northampton St.
Easton, PA

The building had fallen into disrepair and was purchased at a sheriff's auction in 2011. It has since undergone considerable repair to transform this unique building into a modern brewery. From the outside, the charming brick building is striking, with a prominent turret and balcony. Inside, the dark wood, massive bar, tin ceiling tiles, and historic light fixtures provide a memorable ambiance. Reminding customers of the building's past, The **Two Rivers Brewing Company** serves a Pine St. Blonde that is named after the ladies who plied Easton's Red Light District during Prohibition. Along with beers, they serve award winning burgers and my personal favorite item on the menu: poutine.

Church Brew Works, located in Pittsburgh, restored the former St. John the Baptist Roman Catholic Church that had been built in the late 1800s and later expanded in the early 1900s to service the growing immigrant population from Ireland and Scotland. After sitting vacant for many years, Church Brew Works meticulously renovated the building and was able to preserve the stained glass windows and hand-painted ceilings. The detailed work earned them recognition and inclusion on The Pittsburgh History and Landmarks Foundations' List of Historic Landmarks and the prestigious award of "Best Large Brewpub in America" at the Great American Beer Festival. Since opening in 1996, **Church Brew Works** has been a fundamental component to the revitalization of the historic Lawrenceville neighborhood. They serve award winning beers with heavenly-inspired names such as Celestial Gold, Pipe Organ Pale Ale, and Pious Monk Dunkel. The accompanying food strikes a unique balance between Old-World German cuisine and traditional Pittsburgh favorites, served to patrons sitting in the original church pews.

Church Brew Works

Date of Visit: _____

Sampled _____

_____

Comments: _____

_____

The Brewerie at Union Station in Erie, Pennsylvania celebrates the great Industrial Era of the United States and the expansive transportation networks that carried people westward and mobilized essential natural resources. During the first half of the 1800s, canals were regarded as the most efficient way to transport goods, and the Erie Canal became the most successful water transportation system in the United States, connecting the Great Lakes to the Atlantic Ocean. Despite the canal's success, the development of the railroad become one of the most important inventions of the

Industrial Revolution and brought profound social, economic, and political change.

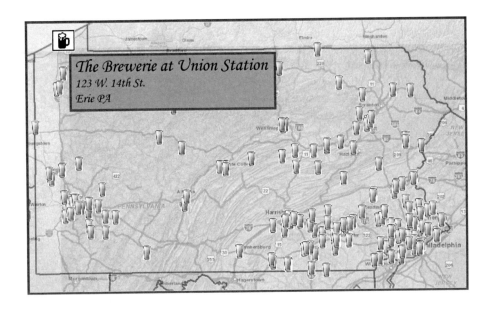

The Brewerie at Union Station
123 W. 14th St.
Erie PA

The Brewerie at Union Station

Date of Visit: _____

Sampled _____

_____

Comments: _____

_____

Alongside the canal, The New York and Erie Rail Road was chartered as early as 1832. During the 1840s and 1850s, Erie experienced a flurry of railroad-building activities with different companies installing converging rail lines.[89] The competing companies used different gauge tracks. Break-of-gauge added delay, cost, and inconvenience to freight and passengers, but usually produced benefits to the local economy because passengers were forced to stop rather than just pass through town. Erie's citizens profited from the break-of-guage as train passengers ate at restaurants, purchased goods from street vendors, and even stayed overnight. In 1953, when a standardized gauge

---

89. Kent, D. H.1948

was laid, the Erie Gauge War erupted, and citizens dismantled tracks and tore down bridges. As part of the dispute's settlement the railroads provided financial support to construct more lines that were eventually acquired by the Pennsylvania Railroad with a 999-year lease.[90]

The first railroad station in Erie was established in 1851 but was replaced in 1866 with a Romanesque Revival-style building. But Erie's population continued to thrive and grow, and in 1927 the city replaced the building with a new Art Deco Union Station. The station was designed by Alfred T. Felheimer, who had been influential in the design of Grand Central Terminal in New York City, and it was the first railroad station in the United States to be built in the Art Deco style. The central design revolved around a large, octagonal rotunda where people could buy tickets, check bags, and purchase news items before descending to a pedestrian tunnel to access the trains' platforms.

The development of the interstate highway, and later commercial airlines, led to the decline of the railroads, and Erie's Union Station closed and fell into disrepair. It was purchased and renovated in 2003 by a logistics and transportation company, who still hold offices in the building. **The Brewerie at Union Station** opened in 2006 and has delightfully recreated the romanticized era of train travel. Seating for the brewpub is in the open rotunda with signs above each of the octagonal walls that in the past would have directed passengers to the checked baggage, tickets, telegraph, and trains. The open space is decorated with old wagons and carts, and wooden barrels provide the base of the tables. The rotating beer menu is written in chalk on an old railroad timetable board. In the summer, customers can order a Rail Way Hefe Weizen and sip the unfiltered, crisp, wheat beer outside in the Trackside Beer Garden and watch passing trains.

Historic preservation of important buildings is alluring but is often costly and restrictive. Furthermore, in a practical sense, old structures over time become functionally obsolete. Adaptive reuse, however, can transform an old space from its original use into one with a new purpose. It can breathe

90. Stowell, D. O.1999

new life into a building, highlight aging assets, maximize architectural and structural potential, and offer an environmentally friendly alternative to new construction. Adaptive reuse allows for flexibility and creativity where designs can pay tribute to history but also recycle a range of products that create an intriguing balance between the past and the modern world. This arduous task has been successfully tackled by numerous breweries throughout the state.

The **Vault Brewing Company** is located in Yardley, a small community along the Delaware River, adjacent to New Jersey. Its natural geography was ideally suited for the development of early industries such as mills and factories. The Delaware Canal opened in 1832 and brought more prosperity to the town. Later, the Reading Railroad, built in 1876, furthered Yardley's expansion. The Yardley National Bank was built in 1889 to support this growing community, and remains today an excellent example of Colonial Revival architecture in a town that has tried to retain its historical integrity. The Yardley Bank closed during the Great Depression, but other money lending institutions utilized the building until 2009 when it was finally left vacant for several years.

The **Vault Brewing Company** purchased and adapted the Yardley National Bank in 2012. Parts of the unique financial characteristics have been careful reused in the modernization of the building. An 8,000-pound vault door from the original building remains a prominent fixture in the brewery, and the bank vault is used as the conditioning cellar. Dark wood, wrought iron fencing, and Victorian lighting fixtures are balanced with shiny copper tank that generate a striking contrast of blending time periods and building use.

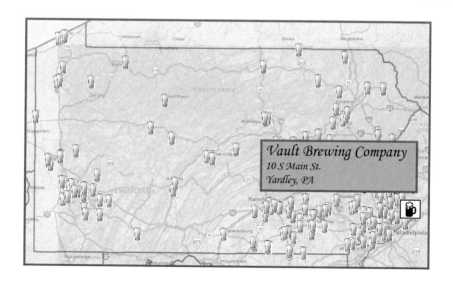

Vault Brewing Company
10 S Main St.
Yardley, PA

Nothing tells a craft brewer's story better than being situated in a town named for a historic tavern. Many early settlers moving west of the Susquehanna River, in today's York County, were farmers. One couple, James and Sara Crosby claimed land holdings in 1785. Following his death, his property was willed to his wife and daughters who were granted a license to operate a tavern in 1804. Sarah Crosby died in 1826 and the tavern operations were taken over by her daughter and son-in-law, who named the tavern Red Lion after his family-owned business in England. Nearly 50 years later, the movement to officially incorporate the town was spearheaded by another tavern owner, Catherine Meyer, who also built a general store, post office, and railroad station. When the borough became officially incorporated in 1880, it was named after the town's first pub: Red Lion.

Red Lion became a thriving stop along the Maryland and Pennsylvania Railroad. It was known for the quality and quantity of cigars the town produced and later became nationally recognized for its lucrative furniture industry. Many of the factories from these industries still remain. A large post office was built in 1935, next to the railroad station, to accommodate this growth. The **Black Cap Brewing Company** renovated this impressive building. With a meticulously landscaped yard and seasonal décor on the wooden doors, the interior atmosphere is reminiscent of early 20th century train travel. Rich wood framed doors and side paneling support the tinted class to the postmaster office where eight rotating taps commonly serve Postmaster's IPA.

Black Cap Brewing Company

Date of Visit: _____

Sampled _____

_____

Comments: _____

_____

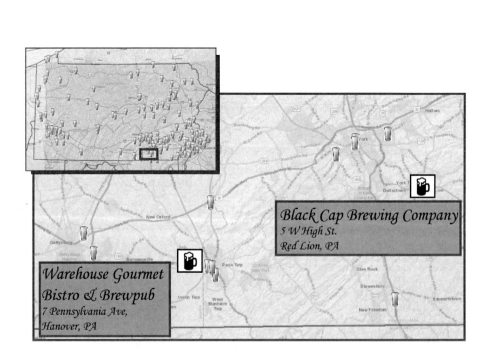

Less than 30 miles away, the town of Hanover has a similar history, with many earlier settlers seeking fertile agricultural lands. Flour, textile, cigar, and furniture manufacturing were prominent industries that later become overshadowed by the famous snack foods enterprises such as Utz Quality Foods, Synder's pretzels, and Martin's potato chips. When the town was originally laid out in 1763, five radiating streets connected Hanover to Abbottstown, Baltimore, Carlisle, Fredrick, and York. The geography of this transportation network made it a prominent center for commerce, which was enhanced even more so by the introduction of the railroad. The town experienced an economic boom between 1870 and 1919, and today Hanover has a designated historic district with many of its buildings dating back to this lucrative era.

The **Warehouse Gourmet Bistro & Brew Pub** is situated adjacent to the railroad tracks in the older, industrial part of Hanover. The brewery proudly promotes the layers of use from the building's 100-year-old history. The menu, photographs, logo designs and beer names celebrate the buildings evolution from cigar factory to ribbon factory, to furniture manufacturer, to frame shop, to storage facility, and art studio. The company started with a catering business that developed into a restaurant, and in 2012, they expanded into the upstairs of the factory building and opened a brew pub.

North Country Brewing, is one of the more eclectic buildings with a fascinating history of reuse. Located in Slippery Rock, Pennsylvania, the building evolved from an original 1805 house and barn to serve as an inn and tavern before becoming a place to manufacture cabinets and coffins, and then serve as a funeral home. In the 1970s the building housed a furniture store. During the renovations and construction of the brewery, which began in 1998, the owners lived in the building and reused as much wood, metal, and stones from the original building and foundation as they could. The reused rocks were transformed into a new patio and a handicap

access ramp. The result is an extremely as-
sorted looking building with unique
character, including a door that reads
"County Morgue, Official Business
Only" and rustic north county decor
of antlers on the wall. In an effort to
be truly sustainable, spent grains are
delivered to the farm and used to feed
the animals raised in nearby fields that
provide much of the food served in the
restaurant. The chicken, beef, bison, herbs,
fruits, and vegetables are locally sourced.

North Country Brewing

Date of Visit: _____

Sampled _____

_____

Comments: _____

_____

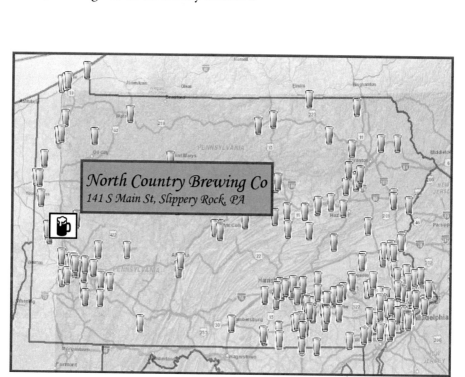

North Country Brewing Co
141 S Main St, Slippery Rock, PA

Union Barrel Works

Date of Visit: _____

Sampled _____

_____

Comments: _____

_____

The list of breweries that promote the reuse of their buildings could go on for chapters. **Union Barrel Works** in Reamstown, for example, previously housed a hardwood store and garment factory. The century-old tin roof and hard wood floors remind visitors of the past. Certainly from a practical vantage point, the use of an older building may be more economical than the expensive rents required in a new strip mall. Renovations can be costly and time consuming, but generally produce a cultural experience that is memorable.

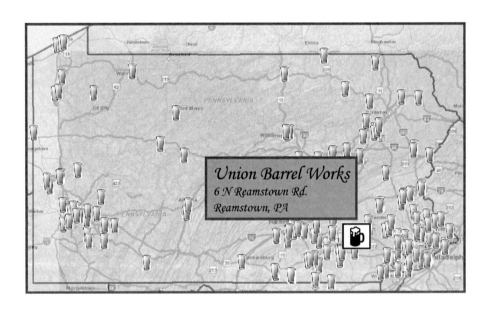

Union Barrel Works
6 N Reamstown Rd.
Reamstown, PA

The promotion of a brewery's hard work of retaining the past while serving the present is often prominently displayed on their menus, framed decorations, and webpages. The promotion of older materials provides the consumer with written, pictorial, and edible experiences that connect them with the past and allows people to LOVE their local beer.

CHAPTER 6

Revitalizing Neighborhoods and Preserving Main Street

Revitalizing neighborhoods by renovating historic buildings and developing new shopping districts, entertainment venues, and cultural attractions has been well-studied.[91] Many attempts, methods, and reasons for neighborhood revitalization have been documented, but none of these initiatives may be as pleasurable as simply enjoying a beer. Local, independent breweries strengthen local economies, reinvent declining neighborhoods, promote sustainable development, and strengthen social history.[92] In some large urban areas and many smaller towns throughout Pennsylvania, the typical "Main Street" has economically declined as large chain stores have developed near interstates and in suburban strip malls. Small, independent businesses can fill vacancies found in traditional downtowns and contribute to creating vibrant neighborhoods by supporting other local businesses and developing a sense of community. Generally, local businesses and local ownership provide numerous advantages, including higher labor and environmental standards, higher economic multipliers, and overall a better success rate than large chain stores.[93]

91. Jayne, Mark. 2006.
92. Feeney, Alison E. 2017.
93. Phillips, Rhonda G and Jay M. Stein. 2013.

Many breweries serve as a corner stone for the community and are central to many neighborhood's economic development. Most of the breweries in Pennsylvania have fully functional restaurants, or at least pub fare or food trucks, that encourage people to gather, sit, and talk. Many craft breweries promote family dining and community interaction where customers can learn, read, and consume their beer surrounded by visual memories of the past. Small towns and neighborhoods across Pennsylvania are being heralded as brewers proudly display local history and promote their connections with the local environment. Craft breweries strengthen local economies, restore unique buildings, and ensure that idiosyncrasies are being passed along in written and verbal form to future generations.[94]

Tattered Flag Brewery and Still Works

Tattered Flag
Brewery and Still Works

Date of Visit: _____

Sampled _____

_____

Comments: _____

_____

Tattered Flag Brewery and Still Works opened in Middletown in 2016 as a veteran owned and operated business with the goal of giving back to the community. Revitalization is not cheap; the town has spent millions improving the streetscape, rehabbing the railroad station, and restoring the town's clock in attempts to attract more businesses.[95] The town has an advantageous location with close proximity to the Amtrak station, Harrisburg International Airport, and Route 283, yet still maintains a small town feel. In addition to the revitalization funds the town invested, **Tattered Flag Brewery and Still Works** contributed a couple million to restore the beautiful 1911 Elk's building in the center of town to create a vibrant business. The brewing system is located in the basement and the distillery and tasting room are located on the ground floor. On the second story, the large space is separated into several rooms, including the original large Elk's bar and open kitchen, a game room, and a separate room for private parties. They use local hops and ingredients whenever possible and support many com-

94. Greco, John. 2016.
95. Miller, Barbara. Oct. 2015

munity events such as the Middletown Fire Department and the local high school Blue Raiders football team. Many of the beer names have a patriotic theme such as War Bonds Stout and ALEways RemAMBER. On a fun note, **Tattered Flag Brewery and Still Works** offers a brew-your-own-beer experience where you and several of your friends can work with the brewers on a professional grade brewing system. After selecting a recipe, participating in as much or as little of the messy work as you prefer, and waiting 2 to 3 weeks, your beer can be enjoyed bottled, canned, or kegged for a special event, and even tapped at the bar.

Several breweries have greatly contributed to revitalizing neighborhoods around older industrial parts of cities. In many instances, these neighborhoods suffered decades of decline, but along with community initiatives, many breweries gleamed new life into older buildings that promote Pennsylvania's brewing history and industrial era.

The **Philadelphia Brewing Company**, for example, is located in the bottling plant of Weisbrod and Hess Oriental Brewing Company. Built in 1885 the company closed in 1938 and the building served as a supermarket warehouse for years. Located in the Kensington neighborhood of Philadelphia — which in the 1980s, 1990s, and early 2000s was notoriously known for its Irish mob and illegal drug sales — the area has recently experienced an influx of working class and young urban professionals. The **Philadelphia Brewing Company** is dedicated to not only restoring the historic brewery but revitalizing the neighborhood. Investment in the neighborhood, impressive murals, community gardens, and even providing lunch for all employees are all part of the company's commitment to social responsibilities. Tours and tastings are open to the public on Saturdays, and on weekdays knowledgeable staff runs a store for both beer and history geeks.

The most notable area of breweries contributing to revitalization of towns can be seen in Pittsburgh. The city, which greatly suffered with the decline of heavy industry in the post-industrial era, has significantly transitioned its economy over the past half-century and has seen growth in education, technology, and health fields.[96] While the entire city and surrounding suburbs has a great collection of beer drinking establishments, the breweries in the Lawrenceville neighbourhood are an integral part of the cultural shopping, dining, and artistic scene. Located northeast of downtown, along the river, many buildings still exhibit signs of the industrial past, which is evident in the two-mile stretch that contains five breweries. One of the newest breweries to the neighbourhood is 11th Hour, or

96. King, C., & Crommelin, L. 2013.

Eleventh Hour
Brewing Co

Date of Visit: _____

Sampled _____

_____

Comments: _____

_____

Eleventh Hour Brewing Co as it is often spelled out, opened early fall 2017. Located in an impressive building that was built in the 1870s as a school for German immigrants, the pillars and brick arches over the doors lead into a family-friendly pub with brick walls lined with wood, exposed duct work, and a bar made from concrete and old doors. Many clocks line the walls, and they are all set to the 11th hour.

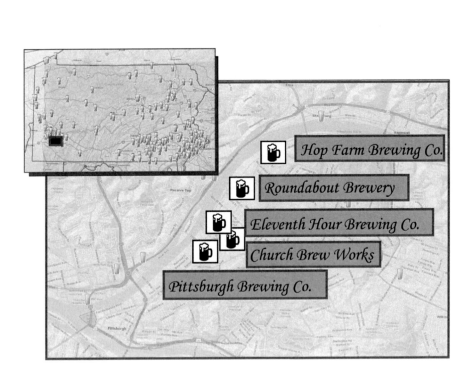

Hop Farm Brewing Co.

Roundabout Brewery

Eleventh Hour Brewing Co.

Church Brew Works

Pittsburgh Brewing Co.

Located just down the street is **Round-about Brewery**, which has been open a few more years. The owners worked in the brewing industry throughout North American and New Zealand and serve an eclectic style of beers. To complement their beers, they partnered with another store to offer New Zealand-style pub pies.

Roundabout Brewery

Date of Visit: _____

Sampled _____

_____

Comments: _____

_____

In the state's capital, the Midtown neighborhood is often cited as a great example of revitalization as manyyoung professionals, artisans, and historic preservationists take advantage of the growing arts, entertainment, and culinary options.[97] **Millworks** is located in Midtown and is part of a restaurant and art gallery that blends the historic past with a forward-thinking sustainable business plan. Using reclaimed lumber and fixtures from the original building, the Stokes Millwork, the brewery is committed to a farm-to-table approach and sources its menu from a local 40-mile radius. At the top of their menu, they express how dining dollars support Pennsylvania's farmers and the brewery's commitment to sustainably managed environmental practices.

Millworks

Date of Visit: _____

Sampled _____

Comments: _____

_____

97. Blumgart, Jake. 5/17/2016.

Just a few blocks away is **Zero Day Brewing Company,** whose owners embrace the neighborhood and are extremely active in community events. The tap room is a delightfully bright, artistic space that has brought new life to a corner building that sat vacant for over two decades and was last used as a plasma donation center. They offer a few snacks, such as a Large Ass Pretzel, but also have a bring-your-own-policy, which is great for people who want to partake in Midtown's growing eclectic restaurants, including the El Salvadorian and Vietnamese restaurants located within one city block. The names of their beers are whimsical and display the characters of the brewers. DeClawed, When

Zero Day
Brewing Company

Date of Visit: _____

Sampled _____

_____

Comments: _____

_____

Did We Get a Dog, ROY G BIV, Cheap Date, and D.T.F. Saison, are just a few of the beers that are worth visiting the brewery to learn first-hand the explanations and stories behind the recipes.

Contributing to Midtown's rebirth as an exciting place to live, work, and shop, **Zero Day Brewing Company** recently opened The Outpost just a few blocks away from their brewery at the Broad Street Market. The Broad Street Market, founded in 1860, is the oldest continuously running farmers market in the country. In recent years, the market has had a noticeable increase in foot traffic as the neighborhood has undergone revitalization.[98] Now with over 40 vendors selling locally grown organic produce, meat, fish, and baked goods, customers can enjoy a pint during daytime hours.

Both breweries in Midtown compliment the other gems in the redevelopment of this neighborhood. Along with the Broad Street Market, the Midtown Cinema is another major draw. It opened in 2001 as an independent theater that shows a range of movies from the classics to recent foreign debuts. Movie goers can bring **Zero Day Brewing Company** beers in and watch a film. Nearby is The Midtown Scholar, an independent bookstore with beautiful wooden shelves and staircases, a coffee shop, and frequent author visits.

---

98. Gleiter, Sue. 4/1/2016

**Mudhook
Brewing Company**

Date of Visit: _____

Sampled _____

_____

Comments: _____

_____

Approximately 30 miles south of Harrisburg, the city of York is redeveloping its central business district and breweries are actively involved. **Mudhook Brewing Company** opened in 2011 in the restored 1883 Central Market House. The owners wanted to be a central figure anchoring the market and seen as one of the premier places to go in downtown York, and thus decided on the name "mudhook," as another term for anchor.

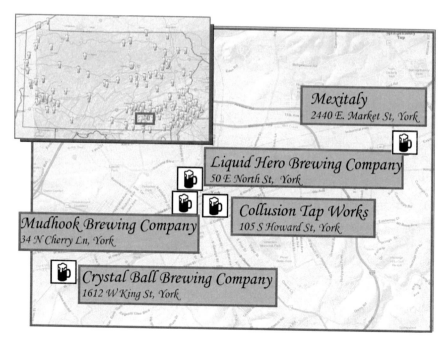

Mexitaly
2440 E. Market St, York

Liquid Hero Brewing Company
50 E North St, York

Collusion Tap Works
105 S Howard St, York

Mudhook Brewing Company
34 N Cherry Ln, York

Crystal Ball Brewing Company
1612 W King St, York

Liquid Hero Brewing Company

Date of Visit: _____

Sampled _____

_____

Comments: _____

_____

Just down the street, **Liquid Hero Brewing Company** purchased an old brick building located across the street from Santander Stadium, home of the York Revolution baseball team. In addition to being a game-time attraction, the brewery offers fun events such as Brewery Bingo, Randall Thursdays, and Yoga Saturdays. The Central Penn Business Journal recognizes York as a haven for microbreweries and commemorates these two breweries for investing money back into the community within the first year of business.[99]

Collusion Tap Works recently opened in 2016 in York and is contributing to the growth of the community. Located in an old Studebaker factory that had been left vacant for years, the surrounding alleyways were deteriorated, dirty, and had dilapidated buildings. Once the brewery moved in and started to renovate, more businesses followed. The alley is now very clean, recently paved with plenty of space for parking, and **Collusion Tap Works** created an outdoor seating area where customers can enjoy beer, play corn hole and appreciate the extensive murals on many of the surrounding buildings.

99. Burkey, Brent. 2011

Inside, **Collusion Tap Works** has brick walls, treated concrete floor, and exposed duct work that gives the large, open space a clean, industrial feel. A wall with glass windows provide views of the brewing systems. The large bar has 24 taps that pour a variety of beer and mead that are produced on the premises, along with local cider and wines. Their head brewer, Jared Barnes, has notable training from the Siebel Institute and Doemens Academy in Munich Germany and work experience at several breweries in the United States. This extensive knowledge and his jovial, inquisitive spirit is what makes this brewery unique. Not only will you find many great beers, but he is always creating new recipes, which makes each visit a fun, exciting experience.

Collusion Tap Works

Date of Visit: _____

Sampled _____

_____

Comments: _____

_____

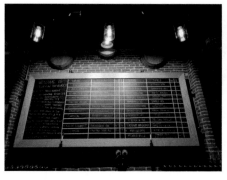

It is not only the large cities that have had economic transitions — many small communities with traditional downtown business districts have declined as suburban development increased, leaving many buildings vacant. Over one-third of Pennsylvania's breweries are located in small communities on typical "Main Streets" and are making significant impacts to their communities and local downtowns.[100]

For example, **Old Forge Brewing Company**, located in the small borough of Danville, advertises that their goal is to make the best beer and food possible and to utilize local businesses, artisans, farmers, and craftsmen whenever possible. The town was once a major transportation hub with canals and railroads, but today it is a peaceful borough located along the North Branch of the Susquehanna River. The historic downtown has an idyllic commercial zone with a coffee shop, running store, bike and ski shop, clothing boutiques, and, of course, a brewery. **Old Forge Brewing Company** is located in a two-story pub. Arched glass windows on the first floor provide views of the brewing system. It is decorated with old mill equipment and all the tables, plates, and iron work is made by local craftsmen. They have an extensive mug program where individuals can purchase personalized hand-crafted mugs made by a local artist. Each tap handle is unique, such as an antler, rail spike, or a wrench that pours beers named Slack Tub Stout, T-Rail Pale Ale, and Plowshare Porter. They also have two beers on a Ye Olde Beer Engine, which is a hand pump that draws the beer from a cask. These beers are unfiltered and unpasteurized and are pulled from the brewing tanks before the carbonation is added. **Old Forge Brewing Company** is active in their community, raising money for local events and charities, including Alex's Lemonade Stand to help fight childhood cancer.

---

100. Feeney, Alison E. 2017.

Old Forge Brewing Company
298 Mill St, PA

Roy Pitz Brewing Company

Date of Visit: _____

Sampled _____

_____

Comments: _____

Many independent businesses, but particularly breweries, have unique settings, menus, and décor option that can be a refreshing divergence from large chain restaurants that serve similar food and mass-produced beers. **Roy Pitz Brewing Company,** in Chambersburg, is an excellent example where they proudly refer to their beer as "Liquid Art," commonly feature local artists and musicians, and support the local economy by actively being involved in community events. Located less than one mile from two consecutive exits off the major I-81 interstate that is lined with chain restaurants, they transformed an early 1900s vacant brick building into a vibrant commercial space filled with local, independent businesses. To celebrate their local connections with the town and its history, one of their flagship beers is Ludwig's Revenge, a dark German-style Rauchbier. It is named in remembrance of the 1864 Civil War event where the Confederate Army raided and burned Chambersburg. Most of the buildings and homes were destroyed including Ludwig's Brewery. **Roy Pitz Brewing Company** not only claims to be located near this original brewery but they use ingredients from Bamberg, Germany where this style of beer originated and, coincidently, is where George Ludwig was born and taught to brew.

Even in suburban areas some breweries have contributed to the reuse of buildings, creating unique dining opportunities rather than cookie-cutter style strip malls and chain restaurants. **Al's of Hampden/ Pizza Boy Brewing Company** in Enola, just west of Harrisburg, has a no-fuss, no-pretense family atmosphere. Located in an industrial park-like setting, they have expanded three times in the past six years. The recent expansion in 2016 focused on an impressive

Al's of Hampden/
Pizza Boy Brewing Company

Date of Visit: _____

Sampled _____

_____

Comments: _____

_____

high-tech BrauKon 35-40 barrel system from Germany, which provides reason enough to revisit this brewery if you remember the small, original store. Along with their brews that you can observe flowing in tubes from the tanks to the tables, **Al's of Hampden/Pizza Boy Brewing Company** has over 100 taps serving beers from great breweries around the state and country.

The spacious building retains that suburban, industrial feel with large doors, open duct work, and metal beams. Large signs instruct visitors on the operations of the brewing system. Customers order food and drinks by numbers at the bar and are expected to pick up their own trays upon leaving. Fun artwork, creative signs, unique bathroom stalls, and great beer and pizza make this one of the liveliest breweries around.

Endless examples exist of breweries revitalizing warehouse districts, supporting local downtowns, and preserving individual characteristics of buildings across the state of Pennsylvania. Craft breweries are reinventing older parts of cities and supporting sustainable development by refurbishing older buildings, maintaining unique architectural characteristics, and stimulating additional economic ventures. It's the personal touch that makes people LOVE their craft beer.

CHAPTER 7

Beer Names and Stories

Have you ever just ordered a drink because of its name? If so, you are not alone. Part of the experience of going to a craft brewery is seeing the array of beers and reading the stories behind how they got their interesting names. A rich literature exists on consumption culture and practices in the United States, where consumption goes beyond the basic needs of people and is instead a cultural phenomenon where people identify commodities with social groups and attach identities.[101] Selecting a beer by its name has become an emotional and sensory experience.

If you are in a mood for raising hell, nothing beats ordering a Purgatory Double IPA, Mischievous Brown Ale, Extra Sinful Bitter, or a Wretched Belgian Strong Golden. **Helltown Brewing** was established in 2011 as a production brewery in Mount Pleasant where you can spend a demonic Friday or Saturday at the brewery with Shawn Gentry, owner and head brewer, who claims the devil made him brew it. Cur-

Helltown Brewing

Date of Visit: _____

Sampled _____

_____

Comments: _____

_____

---

101. Jayne, Mark. 2006.

rently, they are trying like hell to open a tap room in Pittsburgh to let the city dwellers know they are hellbent on making good beer.

Beer has been an important consumption commodity in the United States from early settlement to the present day. The small craft breweries do not have the large marketing expenditures available to the big beer industry, but none the less, they still market and target specific consumers and develop a consumption culture. Many craft breweries focus on people's desire to break away from homogeneity and establish connections with local communities.[102] Many brewers seek out regional lore and emphasize the community's idiosyncrasies with the naming and labeling of their beers. A good beer name should be short, possibly clever, and memorable while considering marketing, labeling, and public perception.[103]

East End Brewing Company in Pittsburgh allows people to sample their beer at their brewery located on Julius Street or their tap room located in the Strip. The brewery offers a great tour for a small fee, which gives visitors a brief history of the brewery, plenty of sampling, and a growler of their favorite beer to go. Many of the beer names reflect the local geography of Pittsburgh's past and present. Smokestack Heritage Porter depicts imagery of the city's past industrial era while Pedal Pale Ale is released every spring as a local charity bicycle event where groups of riders search for a mystery tap location.

Craft breweries promote their location, brand narratives, market their landscape, and tell stories of local history. Many craft breweries name their beers after local leaders, heroes, folklore, and myths to market their beer,[104]

---

102. Flack, Wes. 1997.
103. Carr, Kevin. 2015.
104. Hede, Anne Marie and Torgeir Watne. 2013.

and they display statements of local pride by adorning labels with images of historic miners, loggers, blacksmiths, and captains, along with nostalgic images of trains, horse and buggies, and steamships.[105] Additionally, environmental images such as harvest cycles, mountains, and rivers are commonly used to connect consumers to their beer and local environs. Brewers go to great lengths to create distinctly local themes and images and market narratives of places, sometimes as much as the beer.[106] While ordering and consuming a craft beer, it is hard to ignore the origin, style, and cultural expression intended by the brewer.

Unlike the somewhat dry and traditional history often portrayed in local museums about a town's founding fathers, notable war heroes, or economic resources, breweries proudly promote the more unusual stories, legends, and colloquial events. Craft breweries will promote important political leaders or prominent figures, but are equally eager to represent accounts of illegal activities, unusual sightings, wandering animals, or crazed individuals. The idea of local history and heritage differs greatly from place to place and a town's connection with aliens, monsters, and illegal activities may be as proudly promoted as another town's impact on political independence.[107] While enjoying a beer, patrons learn about intangible culture passed down through stories, local legends, and lore by reading descriptions of food and beverage items. It's that experiential learning that fuels the consumption process and contributes to the consumer's enjoyment of the beer.

Pennsylvania has no shortage of historic events, important people, and wonderful natural resources to have an endless supply of great beer names. As mentioned earlier in this book, **Yards Brewing Company** has an entire Revolutionary flight commemorating important leaders.

105. Schnell, Steven M and Joseph F. Reese. 2014.
106. Schnell, Steven M. 2011
107. Feeney, Alison E. 2015

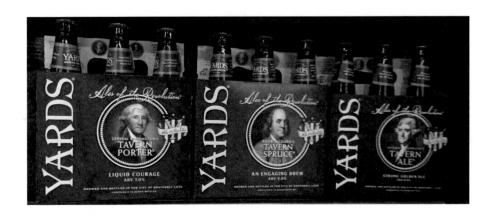

 Erie Brewing Company. Date of Visit: _____ Sampled _____ _____ Comments: _____ _____

Another company that promotes an important leader is the **Erie Brewing Company.** The company was established in the late 1990s and was originally located in downtown Erie. In the early 2000s, the company restructured, and they have recently opened a large state-of-the-art facility near the highway that provides great atmosphere for all. The new facility has large open seating with glass windows with views of the brewing system, several small, semi-private rooms, outdoor seating areas overlooking the brewing system, and even a dog-friendly patio. **Erie Brewing Company** accommodates our need to stay connected and share our experiences with the world by having free Wi-Fi, USB ports, and power outlets throughout the brewery.

For those wanting more knowledge and close up views of the production, **Erie Brewing Company** built Brew House Avenue, which is a glassed walkway with supported steel beams. Etched in the beams are the names of their notable brews. Several of their beer names reflect the town's connections and reliance to the railroad, such as Railbender Ale and DeRalied Ale, but the most exciting name recants local lore with their Mad Anthony's American Pale Ale. It is named for General Anthony Wayne, Brigadier

General of the American Revolutionary War. His story of "madness" from lead poisoning becomes even more interesting when, as lore has it, years after his 1796 death, his son dug up the bones in hopes of relocating him to the family plot, but he drove so fast that many of the bones fell out of the wagon and were strewn across town.

Many breweries tap into the local history and commemorate men of distinction and revolutionary war heroes, but equally as important, many breweries proudly promote local guys. **Full Pint Brewing Company**, for example, is located in North Versailles, just east of Pittsburgh, and added a second location called the Full Pint Wild Side Pub on Butler Street in the Lawrenceville neighborhood. The new pub serves an amazing meat-centric menu sourced from local butchers, along with a few vegetarian options, which pairs well with their flagship beer Gus IPA. **Full Pint Brewing Company** commonly has live music and features local artists, but Gus IPA recognizes a blues player that was closely connected to the brewery since the beginning.

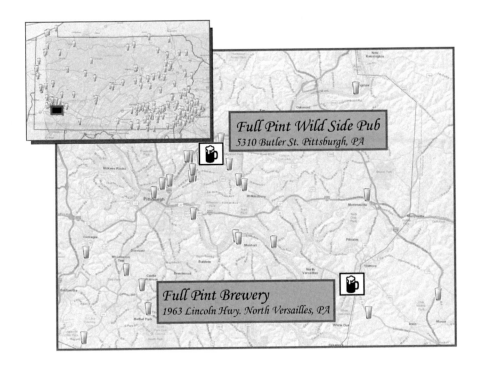

Full Pint Wild Side Pub
5310 Butler St. Pittsburgh, PA

Full Pint Brewery
1963 Lincoln Hwy. North Versailles, PA

You cannot go anywhere in Pennsylvania without recognizing the importance of Penn State and the love of football. Blue and white banners, stickers, flags, clothing, and bandanas adorn houses, cars, people, and dogs throughout the state, so it is not surprising that beer names follow. The **Duquesne Brewing Company** was historically one of the major breweries in Pittsburgh for most of the 1900s. Although it survived Prohibition and became a nationally recognized beer in the 1950s, the company finally dissolved in the 1970s. It was re-established and started producing beer again in Latrobe. In the fall 2015, just in time for football season, the **Duquesne Bottling Company** released Duquesne Lager's Paterno Legacy Series honoring the long-time Penn State football coach and his 409 wins.

Unlike the previous men of action, woman are distinctly absent from marketing in Pennsylvania's craft beers. Marketing campaigns for mass produced beer commonly display female body parts and provocative woman to sell their product; however this is not the case for craft beer. Blonde beer seems to have an obvious naming and some breweries will advertise that with female body parts, but more often than not, women are portrayed in craft beer as dangerous, mysterious, and elusive. **Springhouse Brewing Company** produces The Astounding She-Monster and **The Brewerie at Union Station** honors their resident ghost, a young woman who died in the stairwell with Apparition Amber Ale.

Broken Goblet Brewing

Date of Visit: _____

Sampled _____

_____

Comments: _____

_____

Broken Goblet Brewing is located in Bristol and is known for a number of rotating beers with unique names. They freely admit to liking eclectic, geeky things and are particular nerdy when it comes to Lord of the Rings and Star Wars, so they have beers named A Disturbance in the Force and Angry Troll Amber. However when it comes to including a female named beer, Under Her Spell Sasion pairs well with strong cheese, shellfish, and a safe ride home.

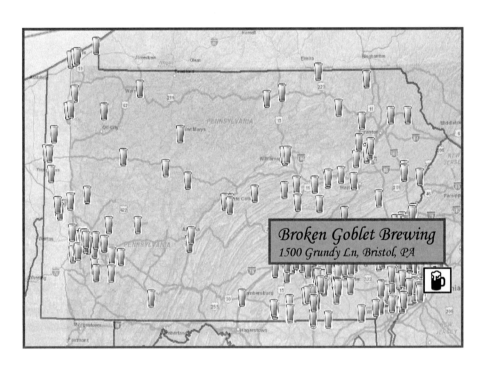

Broken Goblet Brewing
1500 Grundy Ln, Bristol, PA

Historic and industrial images are often used in the naming of beers as many craft breweries portray a romanticized, idealistic past. Railroads were essential transportation in the past to move raw materials, so it is not surprising that they appear in many beer names in Pennsylvania. **Box Car Brewing Company** is located in West Chester. In a beautifully restored brick build-

ing, they have preserved an old stage and decorated with wrought iron, old books, and
leather luggage to create a warm family
atmosphere with nostalgic thoughts of
train travel. Train inspired beer names
such as Passenger Ale, Coal Runner
Stout, and Mr. Pumpcan can be enjoyed with Whistle Down Chili and
Caboose Burgers. In addition to enjoying great beer at their pub, **Box Car
Brewing Company** organizes a Ride the
Rails event where visitors board a 1930s-style
train car at the historic West Chester Railroad station and ride the rails
through idyllic Chester County to Glen Mills for a picnic, live music, and of
course, beer.

**Box Car
Brewing Company**

Date of Visit: _____

Sampled _____
_____
_____

Comments: _____
_____
_____

Many breweries represent the natural resources of the state, particularly the rich agricultural landscape. One great example comes from the **Happy Valley Brewing Company** in State College. This brewery sits in a large barn at the entrance to town. Visitors can eat upstairs in the large open barn, at the long wooden bar, or in the cozy basement with views of the brewing system.

Happy Valley Brewing Company infuses local lore and history throughout the dining experience and informs customers about the history of the brewery and an infamous pig. Legends recall that when the town was incorporated in 1896, they prohibited the grazing of pigs, goats, and horses that previously roamed free in the small village. Ed the Pig refused to stay in his confines and was commonly seen grazing down the roads. He was scheduled to be slaughtered for civil disobedience, but eventually found refuge at the Klinger Farm. Today the renovated barn displays many farm relics and serves Barnstormer IPA and HayDay American Wheat, reflecting the agricultural core on which the community is founded. The brewery logo and tap handles proudly display Ed the Pig. And of course, a craft brewery in State College has to recognize the state's passion for their hometown football team: they serve Tailgater Pale Ale, suggested to be shared with 106,572 of your closest friends.

Happy Valley Brewing Company

Date of Visit: _____

Sampled _____

_____

Comments: _____

_____

Many breweries connect patrons to the local water resources, particularly to their local rivers, and the recreation opportunities they provide. **Barley Creek Brewing Company** is located in the Poconos with easy access to major interstate highways and an outlet shopping center, yet it feels remote as it is nestled amongst trees and located near state forests, a creek, and a ski resort. Their beer names promote the local scenery where they commonly serve Angler Black Larger, Navigator Golden Ale, Slippery Slope, and Antler Brown Ale. **Barley Creek Brewing Company** is one of the state's only breweries that is open for breakfast. They are proud to serve eggs from grass-fed, pasture-raised, antibiotic-free hens and hand-cut bacon from farm-raised, grass-fed cows — it's called Morning Toast for a reason. Making historic connections with earlier farmers and soldiers who commonly started their day with hard cider, whiskey, or bitters, they want to keep that rich tradition going and invite you to raise a glass for breakfast.

Barley Creek Brewing Company

Date of Visit: _____

Sampled _____

_____

Comments: _____

_____

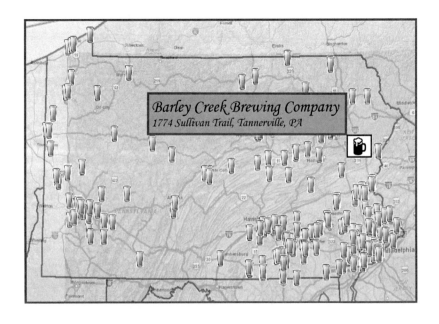

Two breweries located more in the center of the state portray strong environmental images on their website and adorn the drinking establishment with décor that strongly reflects the natural resources in the state. **Bullfrog Brewery** in Williamsport claims that you will love "having a frog in your throat at hoppy hour" and many other amphibian-oriented play on words adorn the location, which is appropriate for their location along the Susquehanna River. This thriving small town has a range of cultural and physical attractions, and Bullfrog Brewery contributes to the vibrant community. The wood-carved tap handles each identify the unique beer names, such as Bully Maker Double IPA, Green Fuzz, and Ghost Frog. One of their flagship beers is called Edgar IPA, named after Edgar Allen Poe's short story Hop Frog.

Similarly, **Elk Creek Café & Aleworks** has a logo claiming they are "rooted in Millheim, PA" with roots extending down to support a pint glass growing in a rolling hilled valley. Their menu proudly promotes use of neighboring farms and vineyards to source their food, and they serve beers with natural, environmentally inspired names. Great Blue Heron Pale Ale, Brookie Brown Trout Ale, Double Rainbow IPA, and Poe Paddy Porter, named for the nearby State Park known for great trout fishing, are commonly found on tap.

Breweries are not afraid to advertise the more urban part of the state as well. **Neshaminy Creek Brewing Company** is located in Croydon and proudly states that they use Neshaminy Creek water in all their beer. The creek borders Bucks County and flows through northern Philadelphia suburbs before entering the Delaware River. Many of their beers reflect the physical geography and have graphic designs that depict the river on their labels. Although they have many great beers, they have won notable awards for their Croydon is Burning Bamberg-style lager in 2016 and 2017 from the Great American Beer Festival.

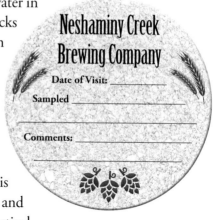

A chapter on beer names could go on for pages. Pennsylvania has many creative brewers who love to connect with their community, reinforce cultural heritage, and actively create a sense of place. Creating and naming something enhances recognition of a place, object, or event, and in modern literate societies, the written text has largely supplanted storytelling. Breweries, however, reinforce, embellish, and embrace local folklore to actively maintain the local culture. A sense of community comes from learning about the obscure; it is an overt statement of local pride. So enjoy reading the description of the names of your beer, ask your brewer for the full story, and don't be shy to ask for the most obscure name on the tap list. You will probably learn something about the town, its history, and the local lore that the community LOVES to share through a BEER.

CHAPTER 8

Ale Trails and Pub Crawls

Most of Pennsylvania's breweries have a warm, neighborly atmosphere where friends and family can gather to recap the day's activities; hikers and bikers can relax after an outing; or out-of-towners can meet the brewers, staff, and locals and leave feeling like they have made a new friend. Breweries provide the perfect community space where food and drink pairs well with all the wonderful activities Pennsylvania has to offer. Whether you are enjoying a staycation, traveling through the state, hiking the mountains, touring museums, visiting historically important battlefields, biking amazing trails, collecting antiques, or strolling through car shows, many breweries have uniquely named beers that pairs perfectly with local activities. Even the Pope enjoyed some Pennsylvania beer on his last visit to the United States.

Tourism is one of the fastest-growing businesses in the world, and many tourists, particularly Americans, enjoy traveling to places of food and beverage production to seek out new palate experiences.[108] Travelers journey to farms, vineyards, breweries, festivals, trade shows, and farmers markets to indulge their palate while immersing themselves in local traditions, cultures, economies, and communities. This experiential consumption creates enjoyment for the traveler that extends past the basic intake of food and drink to include education at the source of production, while travelers also glean pleasure from the regional environs, recreational activities, and cultural lifestyles.[109]

One brewery that is nationally recognized and appears on most travel blogs, tourist guides, and suggested visits to Pennsylvania is **Tröegs Brewing Company.** Founded in 1996 in Harrisburg, they relocated and renovated a building that had originally been built for Hershey Direct Shipping

108. Veeck, Che, and Veeck 2006
109. Hojman and Hunter-Jones 2012

as part of the Hershey Company but had sat vacant for several years. In their new Hershey location that opened in 2011, the abundant space has created a vibrant and thriving establishment where visitors can enjoy a collection of well-known favorites, such as Troegenator, Perpetual IPA, Dream Weaver, and Java Head, along with a creative mix of seasonal beers such as Mad Elf, Blizzard of Hops, and a selection of scratch beers, their small experimental small-batch brews.

Tröegs Brewing Company

Date of Visit: _____

Sampled _____

_____

_____

Comments: _____

_____

Tröegs Brewing Company has created a fun atmosphere that is great for friends and families to gather. The large open space features a long bar with views of the numerous taps in the brewing operation, a snack bar, and a range of indoor and outdoor seating options to accommodate different size parties, moods, and styles. The term snack bar though, is an understatement: **Tröegs Brewing Company** epitomizes the farm-to-table fare with locally sourced ingredients that regularly changes with the seasons. Unlike the common fare you would typically find at a bar, their menu includes unique options such as pig heart pastrami in the charcuteries, mad elf fondue, crispy game hen, duck confit, and octopus tacos.

Visitors of any age can take a self-guided tour without reservations or costs. The path takes visitors through the fermentation hall, quality lab, packaging room, and barrel-aging room. For a longer, more in-depth tour and a minimal fee, visitors 21 years and older can sign up for a guided tour that ends with a sample and souvenir glass to take home. Two maps in the large gift shop shows that people come from all over the United States and Europe to visit this brewery.

The growth in experiential tourism corresponds to the larger farm-to-table movement. Food tourism fuels the notion that there is more to food than food itself, and taken out of its geographic context, it loses its meaning. Part of this growth can be attributed to recent food scares, such as mad cow disease, food recalls, and E. coli outbreaks, along with growing concerns of conflict, terrorism, and global economic meltdowns.[110] Many travelers yearn for a rural, idyllic vacation. This sector of the tourism industry feeds on nostalgia, the vanishing rural past, and many people's affection for the countryside as our society becomes more urbanized.

A great brewery to escape the frenzy of urban cities is in the small town of Bloomsburg. About 100 to 150 miles from major cities like New York, Philadelphia, and Baltimore and easily accessible from I-80, people can escape to the rural countryside surrounded by lush forests and peaceful vistas. **Turkey Hill Brewing Company** is built in the same style as an 1839 Bank Barn that once sat in the same location. Visitors can eat a range of beer-infused menu items and spend a relaxing night next door at The Inn Farmhouse Bed and Breakfast. While taking in beautiful views of the rolling hills and peaceful countryside, visitors can sample their col-

---

110. Schnell 2011

lection of beers and gain an appreciation for the local history and build a personal connection with the brewer by ordering beers with names such as Fort Wheeler's Stronghold Ale, Bamboozled in Bruges Belgian Blonde Ale, Token to Hoboken Hefeweizen, and Rolling Fog Barley Wine.

Many positive benefits come from experiential tourism. It has been hailed as a vehicle for regional development and strengthens local production.[111] Tourism can potentially raise local incomes while offering consumers local products, education, recreation, and socializing opportunities. The notion of eating locally suggests unique choices produced in an ecologically friendly, sustainable way and implies empowering self-sufficient people and other related businesses such as hotels, gift stores, and gas stations that often benefit from increased revenues.

Breweries, along with wineries, cider mills, and distilleries, are at the forefront of this growing industry and drive economic and social development in rural areas because of the collective lifestyle experiences they offer.[112] Parts of Europe and North America, such as Bordeaux, France; Tuscany, Italy; and Napa Valley, California, are well-established experiential tourist regions, but lesser-known areas are documenting rapid growth in tourism and experiential economies. Not surprisingly, most local, regional, state, and even national official tourism websites have designed "ale trails" to encourage people to visit their local breweries.[113]

Several regions within the state have created ale trails with prizes for those who journey to a set number of establishments. The Philadelphia Craft Beer Trail, Lehigh Valley Ale Trail, Greater Reading Wine and Hops Trail, Hershey-Harrisburg Craft Beer Country Trail, Susquehanna Ale Trail, River Rat Brew Trail, Central Pennsylvania Tasting Trail, Cumberland Country Beer Trail, Mercer County PA Wine and Brew Trail, and Lake Erie Ale Trail are just a few. Reflective of the craft brewing industry, these

---

111. Everett and Slocum 2013
112. Ferreira and Muller 2013
113. Feeney 2017

promotional trails by tourism agencies are constantly adapting, so please check local areas for updated specials, prizes, or incentives.

Pennsylvania has many excellent attractions, so you do not have to travel too far to find a collection of great breweries perfect for a create-your-own ale trail and several towns are emerging as hubs for craft beer if you want to stay in one location and have a pub crawl. For example, a traveler could select from over 50 breweries located within a few miles of the 120-mile driving loop around Rt. 581, Rt. 283, Rt. 30, and Rt. 15. This tour is not only in close proximity to the major cities like Washington, Baltimore, and Philadelphia, but it is surrounded by beautiful countryside with pick-your-own fruit farms, hayrides, and farmers markets, museums, exciting attractions, music venues, and world-renowned restaurants and hotels. This trail passes by notable breweries such as **Tröegs Brewing Company, Lancaster Brewing Company,** and **Appalachian Brewing Company** but for someone who would like to enjoy a rural, peaceful experience in the countryside and gain a personal connection by talking with the brewers, this road trip has many small, local options.

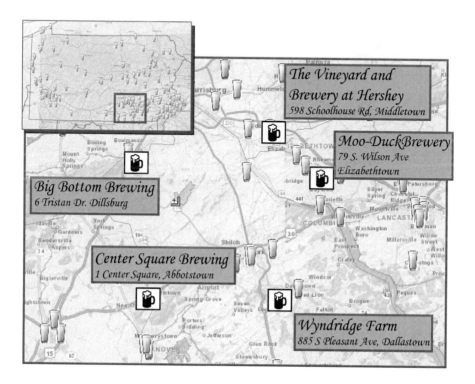

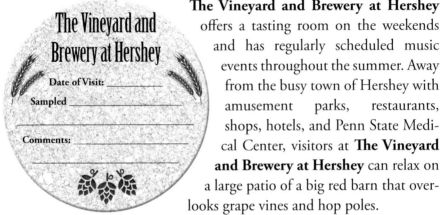

The Vineyard and Brewery at Hershey offers a tasting room on the weekends and has regularly scheduled music events throughout the summer. Away from the busy town of Hershey with amusement parks, restaurants, shops, hotels, and Penn State Medical Center, visitors at **The Vineyard and Brewery at Hershey** can relax on a large patio of a big red barn that overlooks grape vines and hop poles.

A few miles south is the small brewery that specializes in unique, small batch beers. **Moo-Duck Brewery** is located in Elizabethtown, directly across from the Amtrak station, and serves sausages sourced from local meats. This is a small, family run operation where visitors can easily meet the brewer, learn about their process of brewing beer, and gain a better understanding of the local environment. Interestingly, the name of the brewery comes from the owner's participation in an all-star bird watching team. As a sports fan, that is one I had not heard of and is definitely worthy of a trip to hear about the international competition.

Continuing southwest on this road trip, visitors cross the Susquehanna River and drive through the heart of Pennsylvania. Agriculture is the state's leading industry and views of beautiful apple and peach orchards are visible from most roads. The state, particularly with guidance from the Department of Agriculture and Penn State Extension, is working on new technologies and innovations to manage productivity, diversify crops, reduce degradation, and assist with emerging industries such as hard ciders. Amongst the beautiful countryside is **Wyndridge Farm**, which sits on a 77-acre plot of land with a renovated barn. The complex has meticulously developed into a premiere dining facility where visitors can enjoy a restaurant, special events venue, a craft brewery, and cider distillery. Both indoor and outdoor seating provide wonderful views to the rolling landscape.

Center Square Brewing

Date of Visit: _____

Sampled _____

_____

Comments: _____

_____

Deep within the center of civil war history and the fruit belt is **Center Square Brewing**, located in the Altland House in Abbottstown. The brewery is located in the basement of a large house founded in 1753. The renovated late Victorian-style house also houses a restaurant and inn. The brewery itself is a small, three-barrel system that utilizes local agriculture whenever possible, and the beer inspires food pairing options served in the grill.

Completing this ale trail, visitors pass through Dillsburg, a typical small town that still hosts a Farmers Fair in the fall, a holiday light decorating contest in the winter, and Civil War reenactments in the historic Dill's Tavern. Dill pickles in all forms, including fried pickles and pickle soup, are on many local menus and the town uniquely drops a dill pickle on New Year's Eve. For visitors who want that personal experience and the ability to meet and talk to local people, Dillsburg has many options. **Big Bottom Brewery** is located in Al's Pizza and Subs of Dillsburg restaurant. Located just off Rt. 15, travelers can stop in for a slice or two of pizza, or even a full hot pizza buffet. The delightful owner, Bob, knows most of his customers on a first name-basis and is more than willing to pull up a chair and chat with newcomers. Along with a great selection of other local craft beers on tap, **Big Bottom Brewery** runs brews on a two-barrel system and employs two brewers who are constantly refining their craft and collaborating with other brewers to ensure a solid supply of the proven favorites, along with experimenting with new flavors and recipes to ensure variety.

Big Bottom Brewery

Date of Visit: _____

Sampled _____
_____
_____

Comments: _____
_____

If you would rather stay in one location rather than drive, a few towns in Pennsylvania offer a great variety of craft breweries for pub crawls. In some of the larger cities, tours are designed to take visitors to craft breweries, educate them on the process, and ensure they have a wonderful time. City Brew Tours provides a safe and informative tour in several cities on the east coast, including two in Pennsylvania. In both Pittsburgh and Philadelphia, a van will take guests to 4 or 5 breweries, sample up to 16 beers, and provide information on the types of beers, the brewing process, and the history and culture of beer. If you would like some exercise and some fun while in Philadelphia, join a Big Red Pedal Tours where 15 people in an opensided bike pedal their way around to various pubs.

Phoenixville, located about 30 miles northeast of Philadelphia, ranks 10th nationally, amongst cities such as Bend, Oregon; Boulder, Colorado; and Seattle, Washington that host a cluster of great breweries within five miles.[114] A traditional steel city, Phoenixville has undergone enormous economic and infrastructure improvements in the past 10 years and has developed a Community Art Center, a Schuylkill River Heritage Center, and Schuylkill Canal Association, all of which have been influential in the construction of the Schuylkill River Trail, the Mural Arts Movement, the renovation of the 1903 Colonial Theater, and implementation of events such as First Fridays, Canal Days, Farmers Markets, and Festivals. Additionally, environmental initiatives improved one of America's most industrial and culturally significant watersheds, so bikers, walkers, runners, bird watchers, canoers and kayakers can enjoy the only operating lock remaining on the Schuylkill River canal.

114. Caroll, Rich. 2017.

Several of the larger, more established breweries, such as **Sly Fox Brewing Company, Iron Hill Brewery and Restaurant**, and **Appalachian Brewing Company**, were an integral part of the community's development. **Sly Fox Brewing Company**, for example, released STR Ale, where $1.00 of every can sold is donated to improve and maintain the 130-mile Schuylkill River Trail, and **Iron Hill Brewery and Restaurant** supports community bike rides where they donate proceeds from happy hours to Communities in Motion. These breweries helped forge a local craft and artesian culture for the town. In the past few years, several more breweries opened in Phoenixville, with a few more planning to open this coming year.

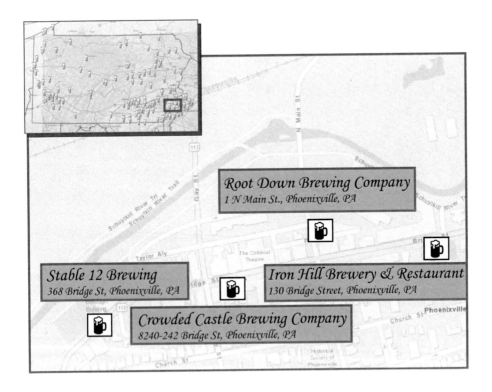

In a one-mile stretch along the commercial center of Bridge Street, Phoenixville has four breweries amongst a collection of distilleries, wine bars, taverns, restaurants, and small businesses that let visitors embrace the local drinks, food, and culture. At the western end of town is **Stable 12 Brewing Company**. The name comes from the initial beginnings: three friends brewing in the family horse farm. The brewery maintains a no-frills atmosphere where customers can focus on the quality of the beer and the company. The bar staff is extremely knowledgeable and personable and sitting inside allows you to learn about their constant rotation of beers, from IPAs to stouts. Across the street is a connector of the SRT trail, and **Stable 12 Brewing Company** has a nice outdoor seating area for those who want to stop after a walk or bike on the trail.

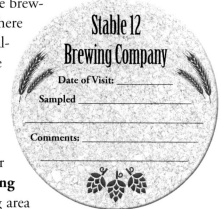

On the other side of town is **Root Down Brewing Company**, located in a large, imposing renovated building. Originally built in 1910 as a theater, the building more recently served as a soda factory before sitting vacant. The **Root Down Brewing Company** tried to retain as much of the original wood, steel, and concrete as possible. The soda factory's large delivery doors and vibrant graffiti creates a hip, industrial feel to the large open space where plenty of tables, a large bar, and a 17-barrel brew house provides enough space for the bustling, hopping, scene that is quickly becoming a foundation of the town.

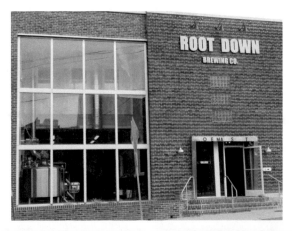

Phoenixville's close proximity to Philadelphia is often attributed to generating the thriving craft beer scene in the suburban community, but what might be surprising is the collection of breweries in the quiet borough of Hanover. A town that once was well connected by railroads and important transportation roads now sits a little off the beaten path. The small town has a fascinating manufacturing and Civil War history, and it retains many architecturally unique buildings that are often overshadowed by today's notable production of snack foods and potato chips. Within a one mile stretch, this town features four craft breweries.

Root Down
Brewing Company

Date of Visit: _____

Sampled _____

_____

Comments: _____

_____

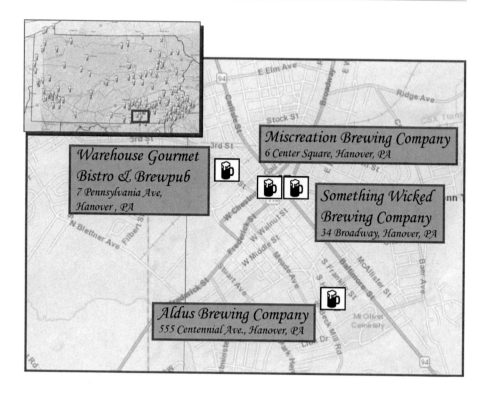

On either side of town the two breweries relive this industrial past. The **Warehouse Gourmet Bistro & Brewpub,** discussed in Chapter 5, is located along the railroad track and proudly promotes the reuse of the old factory building. On the southern side of town is **Aldus Brewing Company**, located in an old Synder's of Hanover pretzel factory. Despite it past use as a factory, the pub is very cozy, and well-known for great service and great beer. It is not uncommon for them to use many local products in their beer, including Pennwood 2017 Harvest Ale, where they used Sunny Brae's fresh Pennsylvania hops and Deer Creek's Pennsylvania malt. The name of the brewery reflects the owner's love of history, and is named after the a 15th-century Italian publisher who

Aldus Brewing Company

Date of Visit: _____

Sampled _____

_____

Comments: _____

is attributed to the survival and printing of many ancient Greek texts, the standardization of italic type, the look of modern commas, and the usage and standardization of the semicolon.

Right in the center of town is **Miscreation Brewing Company** and **Something Wicked Brewing Company**. **Miscreation Brewing Company** has delightfully renovated this building in the center of town with a combination of industrial and vintage-style artwork in a two-story pub. It's impeccably clean, bright space has unique local artwork and serves a variety of sandwiches, salads, and of course, local Utz chips. Literally a few steps down the road is **Something Wicked Brewing Company**, which renovated a vacant 130-year-old building. Serving beers such as Sinful IPA and Wicked Wit, these breweries make Hanover worth the trip.

Miscreation Brewing Company

Date of Visit: _____

Sampled _____

_____

Comments: _____

_____

Pennsylvania has so many small towns and breweries that it is easy to generate your own ale trail or pub crawl, depending upon your time, destination, and personal interest. Throughout the state, travel to breweries and consumption of craft beer is part of a much larger farm-to-table movement. Most breweries in Pennsylvania fully embrace their connections with the local history and geography and engage in sustainable practices that support other local farmers.

Something Wicked Brewing Company

Date of Visit: _____

Sampled _____

_____

Comments: _____

_____

Take time to choose the road less traveled, of which there are plenty in Pennsylvania, enjoy the local environs, engage in a range of experiential activities, talk to a local brewer, and learn to LOVE all there is about Pennsylvania's amazing local characteristics.

CHAPTER 9

Paired Activities

Rather than an entire ale trail or pub crawl, many people just want a beer and a bite to eat after a fun day of activities. Pennsylvania has no end of great destinations with nearby craft breweries to help recap the day's adventure. With great technological advances, in part due to the industrial and agricultural revolutions, over one-third of adults worldwide are inactive.[115] With more people living in urban areas and increased suburban sprawl, there has been a decline in quality green space and an increase in physical and mental health related illnesses.[116] Those who try to remain active often seek monotonous treadmills and stair climbers in gyms or walk in safe, climate controlled malls. Outdoor natural environments can provide health benefits, increase physical activity, reduce stress, restore mental fatigue, and improve self-esteem. The great outdoors should be available for all citizens regardless of age, gender, race, economic status, or physical abilities, and located at a range of distances to provide easy access for all.

One of Pennsylvania's greatest assets is the availability of green spaces that provide playground options to satisfy everyone from the extreme sport adventure-seeker to the urban park bench book reader. Many are free and compliant with the Americans with Disabilities Act (ADA). The abundant

---

115. Hallal, P.C et al. 2012.
116. Gladwell, V.F. et al 2013.

nationally designated parks, scenic trails, and historic sites; and the large number of state and locally managed parks, forests, and trail systems are easily located within close proximity and are easily accessible to much of the state's population. In fact, Pennsylvania boasts 19 National Parks, seven National Heritage Areas, four Wild and Scenic Rivers, five National Trails, over 167 National Natural Landmarks, and one World Heritage Site.[117]

The Appalachian Tail is the longest continuously marked footpath in the world; it runs from Georgia to Maine, with over 200 miles falling within Pennsylvania. The state sees thousands of day hikers, along with zealous journeymen who walk the entire trail. **Zeroday Brewing Company** in Harrisburg is named for the days on the trail where zero miles are logged, often used as a much needed day of rest. One of the owners thru-hiked the Appalachian Trail and knows how important those days are to recover, visit a nearby town to gather supplies, or stop for a beer. Nine breweries, including **Appalachian Brewing Company**, are located within 5 miles of the trail and are perfectly suited to replenish your energy and quench your thirst.

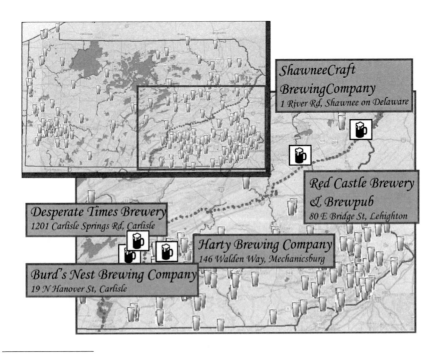

117. United States Department of Interior. 2017

Most thru-hikers enter the state from the south where the Appalachian Trail passes through Caledonia State Park. This is the northern tip of the Blue Ridge Mountains, locally called South Mountain. It is covered in tall pines and has abundant fishing, hiking, boating, and fall foliage viewing opportunities. Long Pine Reservoir sits in the heart of region and is one of the major sources to the Chambersburg water supply and is the main ingredient in all of **Roy Pitz Brewing Company**'s beers. With this pristine water, they create Caledonia Pale Ale, a full-flavored, easy-going ale for days spent hiking, biking, swimming, rock climbing and exploring.

Continuing north, the Appalachian Trail approaches the town of Carlisle filled with relaxing options such as **Molly Pitcher Brewing Company** and **Market Cross Pub & Brewery**. The town recently celebrated the opening of **Burd's Nest Brewing Company** and Grand Illusion Hard Cider. **Burd's Nest Brewing Company** is ideally located near the square and county

Burd's Nest Brewing Company

Date of Visit: _____

Sampled _____

_____

Comments: _____

_____

courthouse in town. Named for the street that one of the owners lived on while attending Shippensburg University, they have craft fully renovated the former Carlisle Arts Learning Center into a spacious brewpub. The tables, walls, and bar uses local wood created by local craftsmen. Large LED screens display sporting events and rotating beer menus.

Desperate Times Brewery is located near the Carlisle's fair grounds, which regularly hosts car shows throughout the year. **Desperate Times Brewery** is one of the town's newer establishments that celebrates the spirit of brew masters during the Prohibition era with beers named Desperate Measures Red IPA and Honest Lawbreaker Oatmeal Stout. They recently received a silver medal at the 2017 Great American Beer Festival for their DTB

Schwartz Beer. They serve authentic Bavarian sausages with names such as Elliot Ness and FDR, but their Oliver Twist, a gigantic warm, soft pretzel is the perfect appetizer to replenish carbohydrates after a day of hiking. This is also a great brewery to stop with a family. They have indoor and outdoor seating, a corner-shelf full of games, and a large chalkboard placed close to the ground for small children.

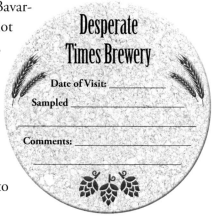

Gardners, Pennsylvania hosts the only museum in the country dedicated to hiking. Located in an old grist mill, The Appalachian Trail Museum is situated close to the halfway point of the entire trail and is viewed as a celebratory accomplishment for many hikers. Nearby is **Harty Brewing Company,** a nano-brewery which produces small batches of beer in a pub located in the Walden community. The taproom serves beer, along with Pennsylvania wines, spirits, ciders, sandwiches, and pretzels. The community development has several unique local businesses and stores, open spaces, and walkways. Additionally, **Harty Brewing Company** has plans to open a larger facility on Market Street in Camp Hill this upcoming year.

Located in the heart of Pennsylvania's Appalachian Trail is Carbon County. Towns like Jim Thorpe, the Lehigh River Gorge, and the Blue Mountains are an outdoor-enthusiast's dream. Nearby Hickory Run State Park has beautiful waterfalls, hiking trails, and a unique boulder field. Enormous sandstone conglomerates forced into a valley by the continual freeze-thaw process during the glacial period have created this spectacular site, which is designated as a National Natural Landmark.

Red Castle Brewery and Brewpub is ideally situated in this region with easy access to great outdoor activities. It serves Polish-American cuisine, including kielbasa, haluski, and pierogis, all of which are a great accompaniment to local beer in a cozy, relaxed atmosphere that pairs well with all the activities the region has to offer.

The Appalachian Trail exits Pennsylvania to the north on the state border, bounded by the Delaware River. Here visitors can enjoy the Delaware Water Gap, a National Recreation Area, and National Wild and Scenic River. **ShawneeCraft Brewing Company**, located on the grounds of the Shawnee Inn and Golf resort along the river, is probably one of the most scenically situated craft breweries in the state. Its beautiful location inspires one to hike, bike, kayak, canoe, ski, or play golf, and of course end the day with an indoor tour of the brewery, a hot pizza and pretzel, live music, and a beer.

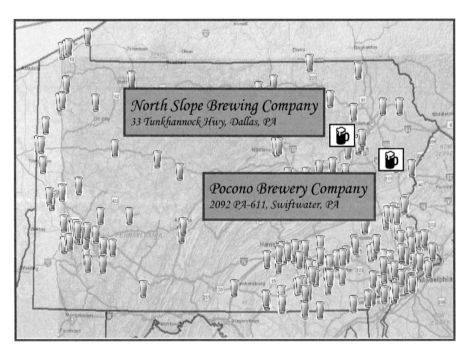

Just a little west of the Appalachian Trail is the Poconos Mountains. The Poconos has been a prime tourist destination for the last 100 years — prior to air conditioning, the higher altitudes, water, and forests provided pleasant summer relief from the hot urban cities like New York, Philadelphia, and Baltimore. The region became a honeymoon destination during World War II, a ski destination in the 1950s, and, more recently, a family-filled destination with large themed water parks and casinos.

Pocono Brewery Company sits in the heart of the region's activities. A family endeavor, the Vitiellos relied on their Italian heritage to renovate a large building and establish a restaurant and brewery that ensures an excellent family experience. They used wood from a local sawmill to cover the walls and ceilings, creating a rustic feel in the dining area that focuses views on an impressive oak wood-burning pizza oven and a beautiful, shiny Criveller brewing system.

Pocono Brewery Company

Date of Visit: _____

Sampled _____

Comments: _____

The father and his sons are commonly seen walking the restaurant to ensure their customers relax and enjoy their food and drinks, but they are equally as passionate when describing their hard work and dedication to creating a building that contributes to the local Poconos culture. A beer menu describes the styles and names of brews that highlight the Pocono's history. Mauch Chunk and 354@35 Drive relive the past, while Pocono Pines Pilsner, 570 Oatmeal Stout, and Pocono Rhubus describe connections to the current economy of the region.

This mountainous region of Northeast Pennsylvania historically brought tourists from all over the East Coast as it became well-known for its endless outdoor recreation opportunities. Back Mountain is known for its rugged beauty, dense tree cover, and scenic lakes. Frances Slocum State Park, located on Back Mountain, has year-round activities focused around a lake created by a flood control structure on the North Branch of the Susquehanna River. Located in Dallas, **North Slope Brewing Company's** name references its location on the mountain. They serve Back Mountain Nut Ale and Dallas Dry Hop, along with a full menu of sandwiches and burgers that will satisfy the entire family's appetite.

North Slope Brewing Company

Date of Visit: _____

Sampled _____

Comments: _____

The majority of the central part of Pennsylvania is a region that is aptly named the Pennsylvania Wilds. It is home to two nationally designated Wild and Scenic Rivers and Allegheny National Forest. They provide endless year-round options to hike, horseback ride, geocache, canoe, kayak, hunt, and fish. Many visitors travel to view the region's two most famous wildlife: Punxsutawney Phil, the weather predicting groundhog, and the wild elk herds that are often seen from designated viewing areas.

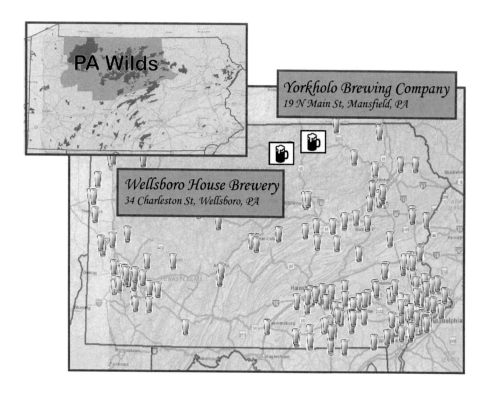

Along the northern border of the state in the Pennsylvania Wilds, The Pennsylvania Grand Canyon stretches nearly 50 miles in length and 1,000 feet deep. Pine Creek runs along the bottom of this gorge and is bounded on either side by Leonard Harrison State Park and Colton Point State Park and is part of Tiago State Forest, near the towns of Mansfield and Wellsboro. Two breweries tap into the natural resources of the state. **Yorkholo Brewing Company** (pronounced

York Hollow), in Mansfield, promotes the state's mountains by serving Endless Mountain Summer Ale, Mountaineer Indian Pale Ale, and Grand Canyon Vanilla Porter. The name comes from the family farm that had to shorten its name based upon stipulation to register cattle. Farming is deeply engrained into the heritage of the brewers who try to utilize local products throughout the season. They serve a Pennsylvania Wilds ale series that uses wild native yeast to spontaneously ferment their beverages, inspired by traditional settlers and their farmhouse ales.

Photos courtesy of Allen Dieterich-Ward

The Wellsboro House Brewery

Date of Visit: _____

Sampled _____

_____

Comments: _____

_____

Nearby, the town of Wellsboro retains a historic charm with authentic gas street lamps and many historic buildings. **The Wellsboro House Brewery** is located in a renovated 1872 house and serves Rail Boss Ale, Coalie Pickman, and Black Indian Pale Ale. Making connections with other Pennsylvania businesses, **The Wellsboro House Brewery** proudly creates Dan Smith's Chocolate Stout, made with local chocolate from a Brookville factory.

Clarion River is a nationally designated Wild and Scenic River and is well-known for its long boating and tubing season. It winds for over 100 miles through dense forests with plenty of public access, so boaters can picnic, camp, hike, bike, fish, or view wildlife along their journey. The river passes through Cook Forest State Park, known for its virgin timber. The Longfellow Trail and Forest Cathedral allow visitors to view the tallest tress in the northeastern United States. **Clarion River Brewing Company**, on Main Street in the small town of Clarion, sits just off the river. The rustic pub serves beers named to celebrate the history and geography of the river: Iron Furnace APA, Clarion River Imperial Stout, 1841, and Leatherwood Brown Ale allow customers to learn more about the region. With connection to Clarion University, part of the Pennsylvania State System of Higher Education, they serve 4.0 G.P.A.

Clarion River Brewing Company

Date of Visit: _____

Sampled _____

_____

Comments: _____

_____

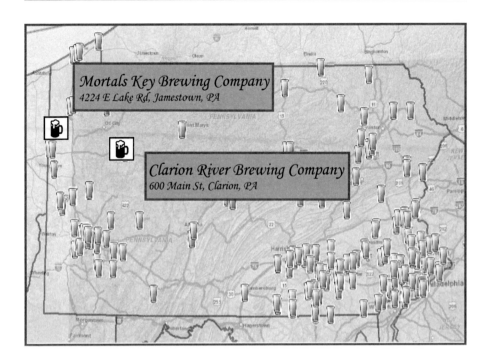

On the western side of the state is Pymatuning State Park. This is one of the most visited state parks in the commonwealth and houses Pennsylvania's largest lake.[118] Not surprisingly, fishing, boating, swimming, and camping are common activities. Along the shores of this lake is **Mortals Key Brewing Company**, a brewery that makes Old World-style small batch, artesian brews. They proudly promote the use of their deep-well spring water as the main ingredient for their beer. Their logo is the Egyptian hieroglyphic symbol for the key of life, and a few hours relaxing in their beautifully renovated tractor barn on a small farm amongst neighboring vineyards and Pymatuning State Park will certainly help anyone relax and enjoy life.

---

118. PA DCNR. 2017.

Along with parks and rivers, Pennsylvania has an enormous number of miles of old rail lines that are transformed into community recreation corridors. Rails-to-Trails Conservancy is a nonprofit organization that supports the development of a nationwide network transforming former railroad lines that allow people of all ages and abilities to walk, run, ride bicycles and horses, and cross-country ski away from traffic. Pennsylvania's dominance in the steel, coal, and iron industries required miles of flat rail lines, which over the years have become abandoned strips of land that have been converted to community recreation trails. Several breweries in Pennsylvania support rail-trails with monetary contributions, donations, naming of beers, and volunteer maintenance. A few breweries have prime real estate along the trails and will refill water bottles, provide great opportunities to rest and refuel, and often keep a tire pump and repair kit on hand.

Rusty Rail Brewing Company

Date of Visit: _____

Sampled _____

_____

Comments: _____

_____

A great destination brewery for a day spent on a trail followed by excellent food and drink is **Rusty Rail Brewing Company**, located in Mifflinburg. The large building was used to manufacture buggies in the early 1900s, billiard tables in the 1940s, and cabinets in the 1950s. The brewery has exceptionally renovated the building, sparing little expense, to ensure guests have an incredible experience. The large dining room on the main floor has exposed brick walls with thick pine beams and glassed views of the 15-barrel brewing system. It is decorated with numerous antiques and rail spikes that are artistically used for tap handles, door handles, and as posts for glass barriers between tables. Upstairs is a game room featuring billiard tables similar to the ones previously manufactured in the building, a safari room, and private banquet area. Large murals portraying the history of Mifflinburg case the stairway and are visible from both the upstairs and downstairs dining area where they serve beers named Rail Spike IPA, Train Wreck Imperial IPA, and a sidetrack series of limited edition beers.

Outside is an impressive patio with an enormous outdoor fireplace that sits along the Buffalo Valley Rail Trail. The paved path can be accessed from **Rusty Rail Brewing Company**'s parking lot, where it turns into a well-maintained crushed stone path and meanders 9 miles towards the town of Lewisburg. Interpretive signs and beautifully landscaped rest stations allow visitors to learn about the history of the area as they ride through tranquil farm fields. I would highly suggest eating at Rusty Rail Brewing Company before an adventure on the Buffalo Valley Rail Trail and return for desert and a Trails End IPA — that makes a great day.

The Lebanon Valley Rail Trail traverses through the heart of Pennsylvania Dutch country with beautiful views of fertile farmland. This well-maintained trail has a crushed-stone path for walkers, runners, and bicyclists that parallels a wood chip and dirt path for equestrians. Interpretive signs celebrate the past 150 years of the Cornwall Furnace National Historic Site and its history relating to mills, canals, railroads, and farming. The trail ends near the village of Mount Gretna with its colorful Victorian cottages. **Mount Gretna Craft Brewery** relives this history at Leeds Corner. In 1924, the corner housed a general store and gas station and was the center of the community. Today they are reinventing the corner neighborhood with a collection of services including Bicycle Outfitters, a store that

Mount Gretna Craft Brewery

Date of Visit: _____

Sampled _____

_____

Comments: _____

_____

provides gear and repairs, and the Red Canoe General Store that houses Sonder Coffee Company and sells food and crafts from local farmers and artists. **Mount Gretna Craft Brewery** creates many beers such as Governor Dick Pale Ale, 1883 Iron Master Stout, and Chiques Creek IPA that commemorate the local history. Proceeds from their Lebanon Valley Rail Trail Ale support the trails maintenance.

A few miles away is the Northwest Lancaster County River Trail that winds for 13 miles along the Susquehanna River. Several miles of this trail is paved, so it is ideally suited for inline skating, wheelchairs, or those who prefer a smooth surface. The southern end of the trail runs through Chickies Rock County Park and an old railroad tunnel to the town of Columbia. The town has several attractions, such as several large antique markets and the Turkey Hill Experience where visitors can learn about dairy farming, make their own ice cream, and indulge in an unlimited number of ice cream samples. However, if you are looking for a craft beer, **Columbia Kettle Works** is

Columbia Kettle Works

Date of Visit: _____

Sampled _____

_____

Comments: _____

_____

located a quarter of a mile from the trail on the main commercial strip of town. With both indoor and outdoor seating, they serve soups, salads, and paninis to accompany their beer, along with local wines and mixed drinks made from local distilleries.

Conshohocken Brewing Company has their production brewery and taproom located adjacent to the Schuylkill River Trail, which runs about 27 miles from Valley Forge to Philadelphia. This is a busy commuter and recreational path and the brewery has numerous bike racks to accommodate the many travelers. People dressed in anything from suits to spandex will stop along the trail to grab a pint and analyze their KOM — because if it's not on Strava it didn't happen. **Conshohocken Brewing Company** also has a brewpub in Bridgeport, a Rec Room in Phoenixville, and a-soon-to-open Town Tap in Havertown.

While the number of miles incorporated and maintained by the Rails-to-Trails program continually increases, some remain small, such as the 1-mile section through downtown Chambersburg that ends at the Pump Track, a dirt terrain with jumps and banked turns. Even a small segment of trail provides opportunities for people to get outside and enjoy a jog, walk, or bike ride by themselves, with friends, or with their furry companions. **GearHouse Brewing Company** was founded by three couples who share their love for bicycling and beer. The brew house is in a renovated industrial building and themed with geared images. Several beer names have bike-related themes, including a number of different beers listed in a Shifting Gears and Single Speed series. They are committed to the revitalization and development of the town, have significantly improved an industrial site, planted a small hop yard, and commonly contribute to charity events. They also created Pump Track Pale Ale, whose proceeds support the expansion of the local bike park at the end of the Rail Trail.

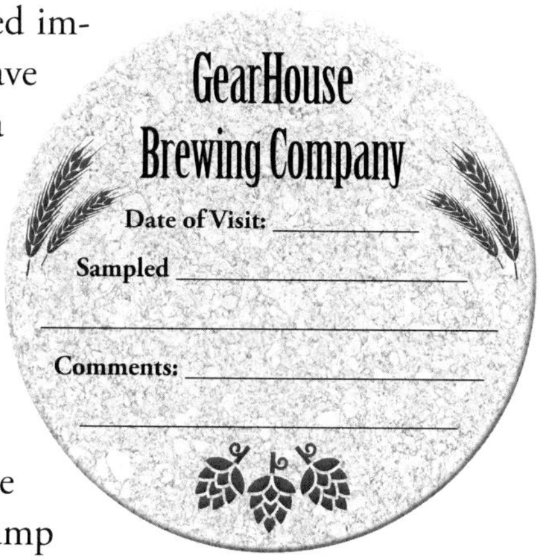

GearHouse Brewing Company

Date of Visit: _____

Sampled _____

_____

Comments: _____

_____

Ever Grain Brewing Company

Date of Visit: _____

Sampled _____

_____

Comments: _____

_____

Not all activities have to be so adventurous, and sometimes the weather does not co-operate, but Pennsylvania breweries keep people active. Many breweries offer yoga. Numerous studies suggest that yoga benefits both physical and mental health and may be as good as or better than other forms of exercise for improving a variety of health-related outcomes.[119] **Ever Grain Brewing Company**, located in Camp Hill, offers free yoga three times a week. Located in an old car dealership along the Carlisle Pike, the brewery offers a large, clean space to clear your mind. The central focus of the brewery is on the enormous wooden bar and the shiny fermentation tanks of the brewing system.

119. Ross, Alyson and Thomas, Sue. 2010.

They have retained the large dealership doors that can be opened in warm weather and creatively built a barn-like facade with a window to order food.

Pennsylvania has many parks, historic sites, natural rives, rail trails, and urban community centers that have an enormous potential to improve physical activity and encourage healthy lifestyles for all their users. Green spaces improve the livability of a place and craft breweries encourage, support, and make connections to Pennsylvania's assets. So start your day with beer yoga, avoid the lunchtime meal at your desk and eat outside, or end a great outdoor day with local food and drink from a craft brewery. I'm not a medical doctor, but I've experienced much of Pennsylvania's outdoor wonders and done my fair share of hours on the stair climber at the gym. I know which provides me with a better feeling of physical health and mental happiness.

CHAPTER 10

Forward Thinking

Beer has been a healthy and nutritious staple since the onset of human civilizations. It has been the impetus for great inventions that have transformed our societies. Early Europeans brought beer to help them survive and settle the New World and in doing so, established Pennsylvania as a leader in world production. Breweries forged themselves as cornerstones of communities that helped shape our political system, and more recently, the mass-produced era of brewing has sponsored sporting events and entertainment venues that provide endless hours of recreation. It is an exciting time for breweries. Large breweries are global giants with the capital and clout to significantly influence and leverage change, and small craft breweries make significant impacts to local communities. Beer and their brewers continue to make improvements to technology, initiate new business ideas, establish social norms, and launch new inventions.

Space Exploration Technology Corporation, commonly known as SpaceX, is a private American aerospace company founded by Elon Musk, the inventor and engineer of Tesla. This leader in electric cars, solar panels, and lithium-ion batteries is working with NASA to launch unmanned spacecraft loaded with supplies to the international space station. SpaceX plans to conduct a reconnaissance mission to Mars in 2022 to establish initial power, mining, and life support infrastructure. They then plan to send a manned crew in 2024 to establish a base civilization.[120] Recent funding for one of the SpaceX launches was sponsored by Budweiser, who sent barley grains for gravity-free research. Budweiser states that information about growth under different environments will further our knowledge about crops here on Earth, particularly with regard to climate change, but they admit space travel is a hot topic, and they want to be the first to brew beer on Mars.[121] The surface gravity on Mars is only 38 percent of that on Earth, so just think about the possibilities for lite beer advertisement.

Modern medicine has detected health benefits from beer. Xanthohumol is a natural product found in hop cones. Although some is lost during the brewing process, different beers retain different quantities of xanthohumol, which has therapeutic properties and is presumed to prevent various cancers, cardiovascular, and other degenerative diseases.[122] Studies in animals

120. SpaceX. 2017.
121. Wattles, Jackie. 2017.
122. Magalhães, P. J. 2009.

have shown that beer prevents osteoporosis, obesity, and type 2-diabetes,[123] while tests on food production indicate meat marinated in beer cuts the number of cancer-causing compounds that form during cooking.[124]

Sustainability is one of the fastest growing and most important fields of the 21st century.[125] It focuses on understanding the interconnections between the economy, society, and the environment, while ensuring an equitable distribution of resources and opportunities. While businesses of all types incorporate sustainable practices, many craft breweries and their customers believe this stewardship is at the forefront and is a top priority of their decisions. In Pennsylvania, numerous breweries have committed to lowering their carbon footprint and ensuring social enrichment for their workers, customers, and communities.

123. Kondo, Keiji. 2004.
124. Viegas, O., Moreira, P. S., and Ferreira, I. M. 2015.
125. Hoalst-Pullen et al., 2014

A few breweries tap into alternative energy sources. **Yards Brewing Company** was the first to harness wind power in 2011. They recently moved to their new, much larger facility but still maintain their resolve to minimally impact the environment. The new location on Spring Garden Street, Philadelphia, offers a much larger dining hall with long wooden tables, a game room with pool tables and shuffleboard, and an upstairs room for private functions. Customers have great views of the enormous production facility, which will nearly double the amount of beer produced from their previously site. Additionally, they plan to add a canning line in the near future. A small gift store allows people to take their favorite beers home with them. Located only a few blocks from major attractions like the Liberty Bell, Independence Hall, and Reading Terminal Market, this move creates a short, easy walk for visitors to access their award-winning beers.

Fegley's Brew Works, founded in 1998, has two locations: Fegley's Bethlehem Brew Works and Fegley's Allentown Brew Works. Both have remained committed to the Lehigh Valley as it has struggled to transition from its previous industrialized steel-based economy. The two locations have full, family-style restaurants where they compost food scraps and operate on solar and wind energy.

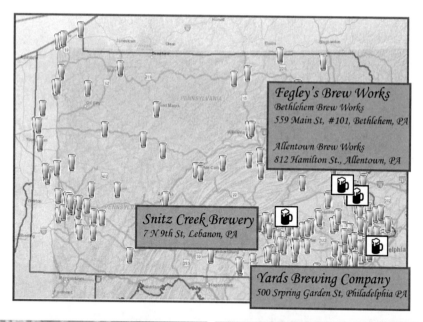

Snitz Creek Brewery is clearly focused on the environment with beer and food names dedicated to the local streams and abundant fish species. Items such as The Shore Lunch, The Keeper, and The Creek Side Burger pair well with Brown Trout Stout, Opening Day IPA, or Linebreaker Double IPA that are poured with fly fishing rod-shaped tap handles. Outside, on the roof above

Snitz Creek Brewery

Date of Visit: _____

Sampled _____

Comments: _____

the beer garden, guests can see first-hand the solar panels that harness the sun's energy, and inside this cozy pub, glassed views of the brewing system have informative signs explaining the brewing process and how their solar panels heat water to 180 degrees to reduce energy and labor costs.

Many breweries recycle their grain and are committed to reducing what goes into their landfills. Most breweries work with local farmers who feed the spent grains to their cattle, pigs, and chickens or compost the organic materials. Other innovative approaches include using spent grains for breads and dog treats.

Most breweries in Pennsylvania are thriving. Within a few short years of operation, many breweries in the state have opened a second taproom, have invested in equipment to help increase their production, and have built small laboratories to help study their beer and control the quality of their product. Additionally, many breweries have expanded their recipes and have ventured into new styles. Fruited IPA, barrel-aged, and sour beers are creating a diversity that customers are excited for and are encouraged to return regularly to sample different options. Many breweries have a strong visual identity and have branded their logo, so as the numerous tasting options increase, a commonality within company's labels and images exists.

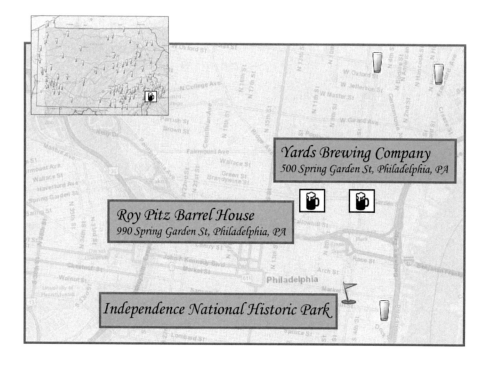

Roy Pitz Barrel House in Philadelphia's emerging Spring Arts neighborhood is an excellent example of the success and prosperity of breweries in the state. As a satellite location removed from their original Chambersburg brewery, they are located a few blocks from the new **Yards Brewing Company**. The building retains its large duct work, exposed brick, steel beams, round pillars, and industrial feel with the addition of an elaborate, vivid painting of the ocean with gigantic octopus tentacles wrapping around two walls. Their signature phrase — "Liquid Art" — merges the line between the artistic setting and the beer. In addition to their flagship beers and seasonal rotations, the **Roy Pitz Barrel House** focuses on barrel-aged and sour beers. Views of the barrels from the tap room help to inform and educate customers on the science behind the wild fermentation process and yeast collection needed to create these unusual, funky, but amazing beers.

Roy Pitz Barrel House

Date of Visit: _____

Sampled _____

_____

Comments: _____

_____

Free Will Brewing Company is another brewery that defines success and provides insight into the growth of the industry. The company has grown over the years with an impressive production and tap room located in Perkasie, and a second tap room in Lahaska. As the company has grown, they consciously work to reduce their carbon footprint, make their production more efficient, quality check and control their product, and have an impressive array of different styles, including an enormous collection of barrel-aged and sour beers.

Free Will Brewing Company

Date of Visit: _____

Sampled _____

_____

Comments: _____

_____

Originally located in the basement of a clothing factory, the production facility now occupies a much larger space on the first floor that is easily seen from the tap room. **Free Will Brewing Company** expands their tap room on the weekends and place chairs and tables amongst the brewing tanks to accommodate the large number of customers. They have invested in cans to increase their distribution and sales, they continuously reinvest in new equipment to help gain consistency in their product, and have built a science lab to conduct analysis. They use eco-friendly carriers for their six-packs and give their spent grain to local farmers. **Free Will Brewing Company** has developed a solid reputation for sour beers and the original basement space its utilized to bottle and store all the barrel-aged and sour beers.

The future is bright for Pennsylvania's breweries. Increasingly, some of the Prohibition-era laws that seemed somewhat illogical are slowly changing. Distribution restrictions are loosening, such as the sale of beer only by the case rather than individual bottles or six-packs. The ability to sell beer, wine, and spirits in the same location, sales at some gas stations and grocery stores, and extended hours of sales, particularly on Sundays, are increasing. The ability for craft breweries to offer beverages other than beer is a great alternative to accommodate consumer preferences. Pennsylvania has a growing number of top quality distilleries, vineyards, and cideries that brewpubs proudly serve in an effort to support other local beverages. Apples are a staple in Pennsylvania's agriculture, so as demand for artesian food and drinks increases it is not surprising that the state now boasts nearly 50 cideries with over $25 million in sales.[126] In an effort to showcase the quality and diversity of products, the state hosted the first Farm Show Cider Competition at the 102nd annual Pennsylvania Farm Show.

Many craft breweries started their journey and business adventure when they fell in love with home brewing. The art and science, the failures and successes, created in garages, kitchens, bathrooms, and backyards throughout the state, often with lifelong friends, turned many passions into careers and gave us the thriving businesses that we enjoy today. The future of new endeavors relies on new people learning and falling in love with the process. Home brewing supply stores, brewing clubs, and educational opportunities are vital to the future of this industry. **Rhone Brew Company** provides supplies and equipment. While they will ship anywhere and are available to answer questions via email or telephone, at their store you can learn to brew with their on-site brewing equipment.

Rhone Brew Company

Date of Visit: _____

Sampled _____

_____

Comments: _____

_____

126. Pennsylvania Pressroom. 2017.

In the transition from homebrewer to professional, many enter competitions. These events are often community sponsored events intended to raise money for charity and provide brewers a chance to gain critical feedback, gauge consumer preferences, and improve their craft in a fun, social setting. Similar to the professional industry, home brewers often create names that reflect their personal history or life experiences. For example, Drop Shot Brewing won awards for Dozer Drool Stout, which reveals the brewers love of tennis and his trusty, furry companion. Of course, receiving monetary awards and bragging rights over fellow friend brewers is an added bonus.

Photo courtesy of Matthew S. Ludwig

Craft breweries in Pennsylvania impact numerous other related industries. Of course all this great beer has to be stored in something, and the American Keg Company in Pottstown is the only stainless steel keg manufacturer in the United States. For over a decade, no kegs were manufactured in the United States, and even today most are imported from Germany or China. American Keg Company produces quality products with much lower shipping costs than imported ones. They currently supply kegs to approximately 100 craft breweries in the state and about 1,000 nationwide. This past year, they have seen an increased use from cideries and wineries, who use their kegs to distribute to bars in an economically and environmentally friendly alternative to bottles.

State sponsored programs and government initiatives have been developed to help promote Pennsylvania's agriculture and craft brewing industry. PA Preferred is a public-private partnership between the Pennsylvania Department of Agriculture and companies throughout the commonwealth to ensure that people support and purchase homegrown and locally made products. PA Preferred Brews is the specific beer-focused component of the program that encourages brewers to use Pennsylvania hops, malt, and fruit and provide qualified brewers with trademarked logo stickers and tap handles to pour and serve their locally sourced ingredients beer.

I witnessed the passion and dedication that many craft brewers invest to build local connections, support Pennsylvania's agriculture, brew amazing beer, and commit to sustainable environmental practices in order to ensure a vibrant future for the craft brewing industry one Tuesday afternoon in a small brewery in Bethlehem. **Bonn Place Brewing Company** epitomizes the story, adventure, hard work, and great love of beer. Sam and Gina Masotto took the plunge from working for others in the food and drink industry to start their own business. Sam emotionally describes his struggles, hard work, and deliberation in the year prior to opening the brewery and how they renovated an old plumbing supply store but feels so fortunate to have the support of the local community and appreciates the guidance and connections he builds with the larger, regional brewing industry.

The small brewery, located in downtown Bethlehem, is clearly recognized on the outside by vivid murals created by Denton Burrows, a New York City artist who earned his Bachelors of Fine Art degree from Lehigh University. Inside, **Bonn Place Brewing Company** has a cozy bar and seating area that is decorated with colorful artwork. Don't let the scrambled array of images and trinkets fool you: Sam's attention to detail extends to every aspect of the building that he has ingrained into his workers. As someone who has several particular, compulsive behaviors, I greatly appreciate Sam's meticulous insistence on cleanliness and order — you will find the corners of your toilet paper folded to a point and the rungs of the bars stools all stored in perpendicular order. This attention to detail proves beneficial to his work. **Bonn Place Brewing Company** was awarded silver and bronze medals for their flagship beers — Mooey in the Ordinary or Special Bitter category and Nemo in the English-style Mild Ale category — at the 2017 Great American Beer Festival.

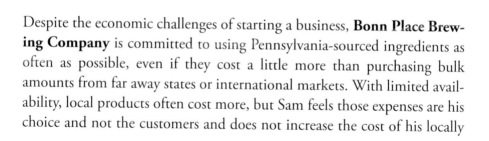

Despite the economic challenges of starting a business, **Bonn Place Brewing Company** is committed to using Pennsylvania-sourced ingredients as often as possible, even if they cost a little more than purchasing bulk amounts from far away states or international markets. With limited availability, local products often cost more, but Sam feels those expenses are his choice and not the customers and does not increase the cost of his locally sourced beers. During my visit to **Bonn Place Brewing Company,** I was fortunate to witness the culmination of a PA Preferred Beer. This 100 percent Pennsylvania-sourced beer started with hops grown on Sunny Brae Hop Farm in Carlisle. The malted barley came from Deer Creek Malthouse; located in Glenn Mills, it is the first commercial malting operation since Prohibition in Pennsylvania. The yeast came from a local homebrewer who grows wild yeast in his backyard and windowsills. Of course, they also used a lot of local Bethlehem water.

Together this collection of farmers, brewers, and craft beer enthusiasts spent the day creating a Pennsylvania pale farmhouse ale where the flavor of the hops and malts are prominently featured. At the end of the day, a farmer picked up the spent grains to feed to his cattle. After a few weeks of fermentation, this PA Preferred registered beer was named after the colloquial saying "Anymore Then Awhile."

Brewing is hard work and a hard science, but they love what they do and have fun doing it. Don't just believe me follow: **Bonn Place Brewing Company** on social media and see the collection of videos they produce that help explain the origins of different beers, demonstrate the brewing process, the collaboration efforts with other brewers, and even poke fun at themselves based on customer reviews. The infectious enthusiasm of Sam and his love of beer carries through to others around him.

Pennsylvania has some well-respected breweries that have withstood the test of time and should provide standards and guidance to establish Pennsylvania as one of the world leaders in quality and quantity of beer production. Many existing breweries are doing well; rarely has a brewery closed in the state. Most are looking to expand, either with a new tap location or with canning operation and distribution of sales. Many new breweries are already set to open. So raise a glass to the past, present, and future of Pennsylvania's brews; it's all for the LOVE of BEER.

References

Baginski, James and Thomas L. Bell. 2011. "Under-Tapped? An analysis of craft brewing in the southern United States," *Southeastern Geographer,* 51(1):165-185.

Baron, Stanly. 1962. *Brewed in America: A History of Beer and Ae in the Unites States.* Little Brown and Company: Boston, MA.

Batzli, Samuel. 2014. "Mapping United States Breweries 1612 to 2011" in Patterson M. and N. Hoalst-Pullen (Eds) *The Geography of Beer: Regions, Environment, and Societies,* Springer-Science: Dordrecht, Netherlands.

Blumgart, Jake. 5/17/2016. "Rebirth in Midtown" retrieved from https://www.keystoneedge.com/2016/05/17/rebirth-in-midtown-harrisburg/

Brewers Association. 2017. "National beer sales and production data," accessed on 10/11/2017 retrieved from https://www.brewersassociation.org/statistics/national-beer-sales-production-data/

Burkey, Brent. 2011. "The business of beer: Beer is not a get-rich-quick endeavor," Central Penn Business Journal. Accessed 10/24/2015. Retrieved from http://www.cpbj.com/article/20111028/FRONTPAGE/111029831/the-business-of-beer-more-than-just-beer.

Caroll, Glenn R. and Anand Swaminatham. 2000. "Why the microbrewery movement? Organization dynamics of resource partitioning in the U.S. Brewery Industry" *American Journal of Sociology,* 106(3): 715-762.

Caroll, Rich. 7/22/2017. "Phoenexiville craft beer boom: Ranks 10th nationally for most breweries," Ornery Ales Production. Retrieved from https://www.orneryales.com/single-post/PhoenixvilleTop10breweryc-ity on 12/15/2017.

Carr, James H and Lisa J Servon. 2009. "Vernacular culture and urban economic development, *Journal of the American Planning Association,* 75(1):28-40.

Cavicchi, Clare Lise. 1988. Beer Brewing at Pennsburg Manor*: History and Technique'*, Pennsburg Manor: Marrisville, PA.

Civil War Photo Album. 2017. "Field hospitals" retrieved from http://www.civilwaralbum.com/gettysburg/hospitals11.htm

Combrune, M. 1804. *The theory and practice of brewing.* Vernor and Hood.

Dineley, M., and Dineley, G. 2000. Neolithic ale: Barley as a source of malt sugars for. *Plants in Neolithic Britain and beyond,* (5):137.

Drake, Dawn. 2009. "Rural architecture as a historical cultural indicator: The case of barns in Monroe County, Pennsylvania," *The Pennsylvania Geographer,* 47(2):90-113.

Ebert, Derrek. 2007. "To brew or not to brew: A brief history of beer in Canada," *Manitoba History,* 54:2-13.

Ensminger, Robert. 1992. *The Pennsylvania Barn: Its Origin, Evolution and Distribution in North America.* John Hopkins Press: Baltimore, MD.

Erickson, Jack. 1987. *Star Spangled Beer: A Guide to America's New Microbreweries and BrewPubs.* Red Brick Press, Reston, VA.

Everett, Sally and Susan L. Slocum. 2013. "Food and tourism: an effective partnership? A UK-based review" *Journal of Sustainable Tourism,* 21(6):789-809.

Felberbaum, Michael. 2015, Feb. 6. "Bud campaign highlights rift between 'big beer', craft" *Boston Globe,* retrieved from http://www.bostonglobe.com/business/2015/02/06/budweiser-campaign-highlights-rift-between-big-beer-craft/FpCjis62CULcQdnLr3dYaI/story.html.

Feeney, Alison E. 2017. "Beer-trail maps and the growth of experiential tourism" *Cartographic Perspectives.* 87: 9-28, doi: 10.14714/CP0.1383.

Feeney, Alison E. 2017. "Cultural heritage, sustainable development, and the impacts of craft breweries in Pennsylvania" *Cities, Culture, and Society.* 9:21-30.

Feeney, Alison E. 2015 "A case for crafty beer names: Bottling the cultural image of Pennsylvania's breweries" *Middle States Geographer.48:20-30.*

Feeney, Alison E. 2015. "The history of beer in Pennsylvania and the current growth of craft breweries," *The Pennsylvania Geographer,* vol. 53, (1):25-43.

Ferreira, Sanette L. and Retha Muller. 2013. "Innovating the wine tourism product: Food-and-wine pairing in Stellenbosch wine routes," *African Journal for Physical Health Education, Recreation and Dance,* September (Supplement 2):72-85.

Flack, Wes. 1997. "American microbreweries and neolocalism: Ale-ing for a sense of place," *Journal of Cultural Geography,* 16(2):37-54.

Filmer, Richard. 1998. *Hops and Hop Picking.* Shire Publications: Buchinghamshire, UK.

Freeman, Amy. Nov. 2016. "The revitalization of Manayunk," retrieved from https://blog.coldwellbanker.com/blog/the-revitalization-of-manayunk/

Gladwell, V. F., Brown, D. K., Wood, C., Sandercock, G. R., and Barton, J. L. 2013. The great outdoors: how a green exercise environment can benefit all. *Extreme physiology & medicine,* 2(1):3.

Gleiter, Sue. 4/1/2016. "New vendors arriving at the Broad Street Market in Harrisburg as it undergoes revitalization," Penn Live, retrieved from http://www.pennlive.com/food/index.ssf/2016/03/broad_street_market_harrisburg.html.

Godlaski, Theodore M. 2011. "Gods of drugs: Osiris of bread and beer," *Substance Use and Misuse,* 44:1451-1456.

Greco, John. 2016. "Craft brews spur development," *Planning* 82(8):8.

Halladay, Jessie. 2012. "Tourist drawn to distilleries, breweries" USA Today, March 19.

Hallal, P. C., Andersen, L. B., Bull, F. C., Guthold, R., Haskell, W., Ekelund, U., and Lancet Physical Activity Series Working Group. 2012. Global physical activity levels: surveillance progress, pitfalls, and prospects. *The lancet,* 380(9838):247-257.

Hayashida, F. M. 2008. Ancient beer and modern brewers: Ethnoarchaeological observations of chicha production in two regions of the North Coast of Peru. *Journal of Anthropological Archaeology,* 27(2):161-174.

Haugland, Jake E. Vest. 2014. "The origins and diaspora of the India Pale Ale" in Patterson M. and N. Hoalst-Pullen (Eds) *The Geography of Beer:*

Regions, Environment, and Societies, Springer-Science: Dordrecht, Netherlands.

Hede, Anne Marie and Torgeir Watne. 2013. "Leveraging the human side of the brand using a sense of place: Case studies of craft breweries," *Journal of Marketing Management,* 29(1-2):209-224.

Hoalst-Pullen, Nancy, Mark W. Patterson, Rebecca Anna Mattord, and Michael D. Vest. 2014. "Sustainability trends in the regional craft beer industry" in Patterson M. and N. Hoalst-Pullen (Eds) *The Geography of Beer: Regions, Environment, and Societies,* Springer-Science: Dordrecht, Netherlands.

Hojman, David E. and Philippa Hunter-Jones. 2012. "Wine tourism: Chilean wine regions and routes," *Journal of Business Research,* 65:13-21.

Homan, Michael M. 2004. "Beer and its drinkers: An ancient Near Eastern love story," *Near Eastern Archaeology,* 67(2):84-95.

Jayne, Mark. 2006. *Cities and Consumption.* Routledge: New York, NY.

Kell, John. 2015. "Reversing evolution," *Fortune,* 171(4): 178-181.

Kent, D. H.1948. The Erie War of the Gauges. *Pennsylvania History: A Journal of Mid-Atlantic Studies,* 15(4):253-275.

King, C., and Crommelin, L. 2013. Surfing the yinzernet: Exploring the complexities of place branding in post-industrial Pittsburgh. *Place Branding and Public Diplomacy,* 9(4):264-278.

Klett, Guy S. 1948. *The Scots-Irish in Pennsylvania.* Pennsylvania Historical Association: Gettysburg, PA.

Kohn, Rita T. 2010. *True Brew: A Guide to Craft Beer in Indiana,* Indiana University Press: Bloomington, IN.

Kondo, Keiji. 2004. Beer and health: preventive effects of beer components on lifestyle-related diseases. *Biofactors,* 22(1-4):303-310.

Kopp, Peter A. 2014. "The global hop: An agricultural overview of the brewer's gold" in Patterson M. and N. Hoalst-Pullen (Eds) *The Geography of Beer: Regions, Environment, and Societies,* Springer-Science: Dordrecht, Netherlands.

Lender, Mark Edward and James Kirby Martin. 1982. *Drinking in America: A History,* The Free Press: New York, NY.

Levin, Aron, Joe Cobbs, Fred Beasley, and Chris Manolis. 2013. "Ad Nauseam? Sports fans' acceptance of commercial messages during televised sporting events" *Sports Marketing Quarterly,* 22(4):193-202.

Magalhães, P. J., Carvalho, D. O., Cruz, J. M., Guido, L. F., and Barros, A. A. 2009. Fundamentals and health benefits of xanthohumol, a natural product derived from hops and beer. *Natural product communications*, 4(5):591-610.

Matthews, Vanessa and Roger M. Pieton. 2014. "Intoxifying gentrification: Brew pubs and the geography of post-industrial heritage," *Urban Geography*, 35(3):337-356.

McAllister, Mathews P. 2001. "Super Bowl advertising as commercial celebration," *The Communication Review*, 3(4): 403-428.

McGovern, Patrick E., Anne P. Underhill, Hui Fang, Fengshi Luan, Gretchen R. Hall, Haiguang Yu, Chen-Shen Wang, Fengshu Cai, Zhijun Zhao, and Gary M. Feinman. 2005. "Chemical Identification and cultural implications of a mixed fermented beverage from late prehistoric China," *Asian Perspectives*, 44(2):249-275.

McGovern, P. E., Zhang, J., Tang, J., Zhang, Z., Hall, G. R., Moreau, R. A., ..and Cheng, G. (2004). Fermented beverages of pre-and proto-historic China. *Proceedings of the National Academy of Sciences of the United States of America*, 101(51):17593-17598.

McLaughlin, Ralph B, Neil Reid, and Michael S. Moore. 2014. "The ubiquity of good taste: A spatial analysis of the craft brewing industry in the United States" in Patterson M. and N. Hoalst-Pullen (Eds) *The Geography of Beer: Regions, Environment, and Societies,* Springer-Science: Dordrecht, Netherlands.

McWilliams, James E. 1998. "Brewing beer in Massachusetts Bay, 1640-1690," *The New England Quarterly,* 71(4):543-569.

Merlino, Kathryn Rogers. 2014. "[Re] Evaluation significance: The environmental and cultural value in older and historic buildings," *The Public Historian*, 36(3):70-85.

Meussdoerffer, F. G. 2009. A comprehensive history of beer brewing. *Handbook of brewing: Processes, technology, markets*, 1-42.

Miller, Barbara. Oct. 2015. "Middletown getting 'pretty comprehensive facelift,'official says" retrieved from http://www.pennlive.com/midstate/index.ssf/2015/10/middletown_getting_pretty_comp.html

Moseley, M. E., Nash, D. J., Williams, P. R., Miranda, A. and Ruales, M. (2005). Burning down the brewery: Establishing and evacuating an ancient imperial colony at Cerro Baúl, Peru. *Proceedings of the Na-*

tional Academy of Sciences of the United States of America, *102*(48):17264-17271.

Moss, Kay and Kathryn Hoffman. 2001. *The Backcountry Housewife: A study of Eighteenth-Century Foods,* Craftsman Printing Company: Charlotte, N.C.

Mowen, A. J., Graefe, A. R., and Graefe, D. A. 2013. Research Brief: Results from the Pennsylvania Craft Beer Enthusiast Study, Unpublished report to: *Brewers of Pennsylvania Association.*

Nilsson, Pia. 2001. "Maps, hops, and war," *Bebyggelsehistorisk Tidskrift,* 61:9-21.

Notte, Jason. 2016. "Opinion: Craft beer is getting its first Super Bowl ad," retrieved from http://www.marketwatch.com/story/craft-beer-is-getting-its-first-super-bowl-ad-2016-01-29 on 3/3/2016.

Patterson M. and N. Hoalst-Pullen. 2014. "Geographies of beer" in Patterson M. and N. Hoalst-Pullen (Eds) *The Geography of Beer: Regions, Environment, and Societies,* Springer-Science: Dordrecht, Netherlands.

Pearce, J. 2002. Food as substance and symbol in the Roman army: a case study from Vindolanda. *BAR INTERNATIONAL SERIES, 1084*(2), 931-944.

Pennsylvania Department of Conservation and Natural Resources. 2017. "Pymatuning State Park" retrieved from http://www.dcnr.pa.gov/StateParks/FindAPark/PymatuningStatePark/Pages/default.aspx on 12/28/2017.

Pennsylvania Pressroom. 11/30/2017. "Homegrown hard cider joins growing list of Pennsylvania agricultural products slated for the 102 Farm Show Competition." Retrieved from http://www.media.pa.gov/pages/Agriculture_details.aspx?newsid=624.

Phillips, Rhonda G and Jay M. Stein. 2013. "An indicator framework for linking historic preservation and community economic development," *Social Indicators Research,* 113: 1-15.

Phillips, Rod. 2014. "The Middle Ages, 1000–1500: The birth of an industry," in *Alcohol: A History*: 66-86. University of North Carolina Press: Chapel Hill.

Reid, Neil, Ralph B. McLaughlin, and Michael S. Moore. 2014. "From yellow fizz to big biz: An American craft beer comes of age" *Focus on Geography,* 57(3):114-125.

Ross, Alyson., and Thomas, Sue. 2010. The health benefits of yoga and exercise: a review of comparison studies. *The journal of alternative and complementary medicine, 16*(1):3-12.

Ross, Michael and Paul Marr. 2008. "Socio-economic conditions on the Pennsylvania frontier: The Germans and Scots-Irish of Cumberland County, 1765-1775," *Middle States Geographer,* 41:1-8.

Schnell, Steven M. 2011. "The local traveler: farming, food, and place in state and provincial tourism guides, 1993-2008" *Journal of Cultural Geography,* 28(2):281-309.

Schnell, Steven M and Joseph F. Reese. 2014. "Microbreweries, Place, and Identity in the United States" in Patterson M. and N. Hoalst-Pullen (Eds) *The Geography of Beer: Regions, Environment, and Societies,* Springer-Science: Dordrecht, Netherlands.

Selinsgrove. 2017. "Historic Selinsgrove" retrieved from http://selinsgrove. net/know-the-grove/historic-selinsgrove/.

Shears, Andrew. 2014. "Local to national and back again: Beer, Wisconsin & scale" in Patterson M. and N. Hoalst-Pullen (Eds) *The Geography of Beer: Regions, Environment, and Societies,* Springer-Science: Dordrecht, Netherlands.

Smith, Greg. 1998. *Beer in America: The early years 1587-1840.* Siris Books: Boulder, CO.

Smith, Vanessa. 2006. "Food fit for the soul of a Pharaoh: The mortuary temple's bakeries and breweries," *Expeditions,* 48(2):27-30.

SpaceX. 2017. "Mission to Mars" retrieved from http://www.spacex. com/mars

Stowell, D. O.1999. *Streets, railroads, and the great strike of 1877.* University of Chicago Press.

United States Department of Agriculture. 2016. "National Hop Report" report released by the National Agricultural Statistics Service, Agricultural Statistic Board, ISSN: 2158-7825.

United States Department of Interior. 2017. "Pennsylvania by the numbers" retrieved from https://www.nps.gov/state/pa/index.htm.

Veeck, Gregory, Deborah Che, and Ann Veeck. 2006. "America's changing farmscape: A study of agricultural tourism in Michigan," *The Professional Geographer,* 58(3):235-248.

Viegas, O., Moreira, P. S., and Ferreira, I. M. 2015. Influence of beer marinades on the reduction of carcinogenic heterocyclic aromatic amines in charcoal-grilled pork meat. *Food Additives & Contaminants: Part A, 32*(3):315-323.

Vorel, Tim. 10/2/2017. "176 of the Best DIPA/Imperial IPAs, Blind-Tasted and Ranked." Paste magazine. Retrieved from https://www.pastemagazine.com/articles/2017/10/176-of-the-best-dipaimperial-ipas-blind-tasted-and.html.

Wagner, Rich. 2012. *Philadelphia Beer: A Heady History of Brewing in the Cradle of Liberty*: History Press: Charleston, S.C.

Wang, Jiajing, Li Liu, Terry Ball, Linjie Yu, Yanging Li, and Fulai Xing. 2016. "Revealing a 5,000-y-old beer recipe in China," Proceedings of the National Academy of Science, 113: 6444-6448.

Wattles, Jackie. 12/11/2017. "SpaceX is sending barley to space for Budweiser research" CNN tech retrieved from http://money.cnn.com/2017/12/11/technology/future/spacex-budweiser/index.html.

Wilcox, Gary B. 2001. "Beer brand advertising and market share in the United States: 1977 to 1998," *International Journal of Advertising, 20*(2):149-168.

Yelker, Rama, Chuck Tomkovick, Ashley Hofer, and Daniel Rozumalski. 2013. "Super Bowl ad likeability: Enduring and emerging predictors," *Journal of Marketing Communications, 19*(1):58-80.